CONTENTS

Figures   vi
Tables   viii
About the Authors   xiii
Acknowledgements   xv
Dedication   xvii
Enterprise Ireland   xix
Foreword *by Prof. Roy Green, NUI, Galway*   xxi
Preface   xxv

1 Technology Transfer in Context   **1**
Introduction   1
HEIs in the Knowledge-based Economy   2
The Triple Helix Model   3
The Impetus for Increased Emphasis on Research   7
  Commercialisation
Business Demand for Research Excellence   9
The Growth of Third-stream Activities: 13
  Benefits & Concerns
Strategically Managing Third-stream Activities: 15
  An Issue of Balance
Implications for HEI Leaders & Administrators   18

2 Organisational Structures for Technology Transfer   **23**
Introduction   23
Technology Transfer Offices   23
The Evolution of TTO Structures   27
Varieties of Organisational Structure   32
The Lessons for HEIs   42

3 Technology Transfer Roles, Activities &   **43**
Responsibilities
Introduction   43
Historical Context   43
Roles & Activities of TTOs   44
Responsibilities of TTOs   51

TTO Staffing & Developing Expertise   56
Key Issues for HEIs   60

**4 Mechanisms for Technology Transfer: 63
    Patents, Licensing & Company Formation**
Introduction   63
Some Theoretical Debates: Research   64
   Commercialisation & Technology Transfer
HEI - Industry Collaboration   65
Commercialisation of Research: Some Empirical   69
   Evidence
Technology Transfer   72
Licensing: Activities & Practices   84
Company Formation: Start-ups & Spin-offs   92
Other Technology Transfer Mechanisms   101
Soft Methods of Technology Transfer   103
Implications for HEIs   107

**5 Developing Technology Transfer: 109
    Stimulants & Barriers**
Introduction   109
Stimulants to Technology Transfer   110
Barriers to Commercialisation   118
Evaluating Technology Transfer Activities   127
Measuring Technology Transfer Activities: 131
   Output & Impact Indicators
Measuring Technology Transfer Activities: 136
   Activity Indicators
Implications for HEIs   140

**6 Varieties of Excellence: 143
    Best Practice Cases in Technology Transfer**
Introduction   143
TTOs & Research Commercialisation in the US   144
Technology Transfer & Commercialisation in Israel   155
Research Commercialisation in Finland   164
Key Lessons from Varieties of Excellence Cases   170

7 A Strategic Approach to Technology Transfer   173
Introduction   173
Key Issue 1: Strategic Management of Technology   176
   Transfer
Key Issue 2: Technology Transfer Activities   179
Key Issue 3: Organisational Structure   187
Key Issue 4: Staffing, Skills & Resources   196

Key Issue 5: Policy & Procedures   201
Key Issue 6: Mechanisms for Technology Transfer   212
Key Issue 7: Evaluating Technology Transfer   216
Lessons & Key Issues for HEIs   223

8   **Making Technology Transfer a Reality**   **229**
Introduction   229
Cultural Ethos for Commercialisation   229
Researcher Motivation   233
Lessons for HEI Leaders & Administrators   239
Making Technology Transfer a Reality in Ireland   239
Towards a Strategic Approach to Technology Transfer   245
  in Ireland

Appendices
1   Acronyms & Abbreviations   249
2   Useful Sources of Information   250
3   Bibliography   252
4   Further Reading   263

Index   265

FIGURES

1.1   Sources of Competitive Advantage in the Irish   3
       Knowledge-based Economy
1.2   R&D Expenditure as a Proportion of GDP, by Source   10
1.3   R&D Expenditure by Sector, 2001   11
1.4   Researchers per '000 Labour Force, 1999   17

2.1   UK Technology Transfer Offices   24
2.2   John Hopkins University's Organisational Structure   37
2.3   Pennsylvania State University's Organisational   38
       Structure
2.4   Duke University's Organisational Structure   40

3.1   TTOs Focusing on Technology Transfer, by Country   47

4.1   The Process of University Technology Transfer   73
4.2   Patents Assigned to US Universities, 1969-1999   80
4.3   Invention Disclosures Received by US Universities,   82
       2002
4.4   UK: Licences & Options Executed, 2002   86
4.5   Start-up Companies Formed by US Universities,   93
       2002
4.6   Spin-outs Created & Licence Income, 2002   101

5.1   Structure of the Chapter   110
5.2   Technology Transfer Programme Start Dates at US   115
       Universities
5.3   Barriers Identified by University Scientists   126
5.4   Barriers Identified by Managers   127
5.5   Barriers Identified by TTO Managers   127
5.6   Molas-Gollart *et al.*'s (2002) Framework for   137
       Analysing Activities
5.7   The Contingency Effectiveness Model of Technology   140
       Transfer

6.1   Disclosures at the University of California, 1976-1989   150
6.2   The Commercialisation Process at TRDF   159

6.3 The Finnish Innovation System   165

7.1 Developing a Strategic Approach to Technology   174
        Transfer
7.2 Pasteur's Quadrant   181
7.3 TTO Structure: Specialist Activities   190
7.4 TTO Structure: The Case Management Approach   191
7.5 Structural Changes Required in TTOs   191

TABLES

1.1 Expertise in Technology: Product & Service   3
         Development

1.2 BERD as a Percentage of Output by Sector, 2001   11

1.3 Measures of Performance of the Science Base   17

1.4 Re-conceptualising the University: 20
         Characteristics of Established Research Universities
         & New University Models

2.1 Interface Structures: YISSUM at the Hebrew   27
         University of Jerusalem

2.2 Regional Partnerships for Managing Intellectual   30
         Property

2.3 Guidelines for Developing an Appropriate   31
         Organisational Structure for TTOs

2.4 Competencies & Characteristics of Alternative   34
         Organisational Structures

2.5 A Comparative Summary of Three Universities   36

2.6 TTO Performance and Organisational Structures   41

3.1 Managing Technology Transfer at MIT   46

3.2 Aspects of Relationships/Networks in University- 49
         Industry Technology Transfer

3.3 The Roles & Activities of TTOs   50

3.4 The Roles & Responsibilities of TTOs: 52
         Sample Mission Statements

3.5 Divergent Priorities: Ranking the Importance of   53
         University Technology Transfer Outcomes by TTO
         Officers, Administrators & Faculty

3.6 Revealed Priorities: Percentage of Time on the Job that   54
         Public & Private University TTO Officers Spend on
         Various Activities, 1999

3.7   Characteristics of University Technology Transfer   55
         Stakeholders

3.8   Employment at TTOs at US Universities, 2000   56

3.9   Essential Characteristics of Professional Staff   59
         Employed by TTOs

3.10 Functions/Activities of Technology Transfer   60
         Managers/Directors

4.1   Encouraging University-Industry Collaboration: 66
         Matching Services

4.2   TTO Research Commercialisation/Technology   68
         Transfer Mechanisms

4.3   Success Factors for Commercialisation & Technology   69
         Transfer Activities

4.4   Technology Transfer: Stage Activity Outcomes   74

4.5   Summary of Empirical Research on Technology   75
         Transfer

4.6   European Patent Office Applications & Grants: 79
         Ireland & Finland

4.7   Conditions Influencing the Probability that TTO will   82
         Patent

4.8   Proof of Concept Fund: Scotland   83

4.9   Proof of Concept: Enterprise Ireland   84

4.10 OECD: Licences *Per Annum*   86

4.11 Technology Licensing: USA   87

4.12 Licensing & Exploitation Companies: ISIS at Oxford   88

4.13 Equity Preferences Reported by TTO Officers, 1999   89

4.14 Licensing Revenues & Patents by Academic Field,   90
         1999

4.15 Key Indicators of Earnings by Field of Technology,   91
         1999

4.16 Spin-offs & Start-ups per TTO Reported in the Last   94
         Year

4.17 University of Louvain La Neuve   95

4.18 Warwick University: Institutional Equity Policy   98

4.19 Twente University: The Support Package   99

4.20 Case Study: Edinburgh Crystal, Wolverhampton   103
         University & Edinburgh College of Art

4.21 Connect Midlands   104

4.22   The Teaching Company Scheme in the UK's   105
           Knowledge Transfer Partnerships
4.23   Entrepreneurship Course: USA   106
4.24   Consultancy Case Study: US Policies – MIT   107

5.1   Case Study: The Cambridge-MIT Institute   114
5.2   Cultural Differences: Universities *vs* Industry   124
5.3   Quantitative Measures for Technology Transfer   132
5.4   Key Technology Transfer Indicators   134
5.5   Measuring Technology Transfer   135
5.6   Outputs of University-Industry Technology Transfer,   139
           as Identified by Interviewees

6.1   Licensing Revenues Collected by Universities &   147
           Colleges in FY 2001
6.2   Top Five US Universities Receiving the Most Patents   147
           for Innovation during 2002
6.3   Principles Framework at the University of California 153
6.4   Criteria Used to Assess Invention Disclosures by   157
           TTOs in Israel
6.5   Perceived Success Factors for Technology Transfer   158
6.6   Cases in Regional Co-operation & Company Models   167
6.7   Cases in University-Industry Interaction in Finland   169

7.1   Elements of the Strategic Plan for Technology   178
           Transfer at Georgetown University
7.2   TTO Outreach & Educational Activities   186
7.3   Guidelines for Developing an Appropriate   189
           Organisational Structure for TTOs
7.4   A Co-operative Model of TTOs   194
7.5   Guidelines for Developing Co-operative   196
           Regional/National Structures
7.6   Guidelines to Facilitate the Optimal Recruitment of   197
           TTO Staff
7.7   Intellectual Property Protocol: Main Features   207
7.8   Intellectual Property Rights Guidelines   210
7.9   Potential Indicators to Measure TTO Activities   219

7.10   Developing a Strategic Approach to Technology   224
           Transfer: Key Issues & Facilitating Factors

8.1   The Entrepreneurship Centre at Imperial College,   232
           London
8.2   Factors Facilitating a Change in Mind-set   233
8.3   Creating a Culture & Ethos of Commercialisation   240
8.4   Overview of Connect Midlands   244

ABOUT THE AUTHORS

JAMES CUNNINGHAM (BBS, MBS, PhD) is a college lecturer in strategic management in the Department of Management, a research fellow for the Centre for Innovation and Structural Change (CISC) and Executive MBA Programme Director, J.E. Cairnes Graduate School of Business & Public Policy at National University of Ireland, Galway. Prior to joining NUI Galway, he lectured in the Department of Business Administration at University College Dublin and worked as a strategy consultant. His research interests encompass three areas: strategy practice, strategy and the environment and entrepreneurship and technology transfer. His research has merited national and international distinction. James was awarded the Lord Edward Memorial Bursary to pursue his PhD studies. James won the public sector category of the Irish Case Writing Competition and the European Case Writing Award organised by the European Foundation for Management Development for a case study on the Reform of the Irish Ports. In 2004, he won a case writing competition organised by Enterprise Ireland and the Irish Academy of Management. Based on his PhD research, he won the best paper awards at the British Academy of Management and the Irish Academy of Management and has published his research in a number of leading journals, including *Business Strategy Review*. In 2005, his co-authored paper, 'In Search of the Strategist', won the best paper award for the Strategy as Practice track at the British Academy of Management held at Oxford University. He co-authored *Enterprise in Action*,[1] now in its second edition and has completed commissioned reports for Udáras na Gaeltachta, Forfás and ICSTI and CISC. In addition, James has made a number of invited presentations on the subject of technology transfer, intellectual property, entrepreneurship and strategy practice.

BRIAN HARNEY (BA, MBS) has a first class honours BA degree from the University of Dublin, Trinity College and a first class honours MBS (Corporate Strategy & Human Resource Management) from the National University of Ireland, Galway. In 2004, he received the IMI's Sir Charles Harvey Medal as one of the most outstanding graduates of

---

[1] Cunningham & O'Gorman (2001).

a postgraduate business degree in Ireland. Brian's MBS thesis examined the determinants of HRM in small and medium-sized enterprises. He has presented papers based on this research at national and international conferences, including the Industrial Relations Research Association Conference (2005), and the Irish Academy of Management Conference (2004), where he received a joint best paper award in the HRM track. Brian has published in leading HR journals, including the *Human Resource Management Journal* and *Advances in Industrial & Labor Relations*. Brian's other main research interests include strategy as practice and university technology transfer. In 2005, a paper he co-authored, 'In Search of the Strategist', was awarded the best paper in the Strategy as Practice track at the British Academy of Management. He is also co-author of a comprehensive review of university technology transfer, the 'Varieties of Excellence Report', commissioned by CISC, and has a paper based on this work forthcoming in the *Irish Journal of Management*. Brian lectures in strategy and HRM at NUI Galway, and has work experience in HR and strategy consulting. He is currently pursuing his PhD, funded by a CISC scholarship, at the Judge Business School, University of Cambridge, where he is also the recipient of a Cambridge European Trust Bursary and is a Fellow of the Cambridge European Society.

ACKNOWLEDGEMENTS

We would like to thank sincerely a number of people for helping us prepare this book – in particular, the Registrar's Office, NUI Galway, and Professor Roy Green, former Dean of Commerce and Chairman of the Technology Transfer Review Committee at NUI Galway, for his energy, enthusiasm and support.

We would also like to thank Dr Aidan Kane, Director of the Centre for Innovation & Structural Change, NUI Galway, for providing us with the necessary support and resources, and Angela Sice, at the Centre for Innovation & Structural Change, and Nuala Donohue, at the Department of Management, NUI Galway, for providing us with exemplary administrative assistance.

A special word of thanks to Eimear Harney for reviewing and collating material and attempting to make sense of our notes.

We would also like to acknowledge the assistance and advice provided by Will Geoghegan, Brian Clarke and Dr Paul Ryan.

We owe a huge debt to Brian O'Kane of Oak Tree Press for realising the value of this project and for providing the encouragement and support that made the publication task all the easier.

Finally, we gratefully acknowledge the grant contributions of the *NUI Galway Grant-in-Aid Publication Scheme* and the *National University of Ireland Grant-in-Aid Publication Scheme,* and the support of Enterprise Ireland.

In loving memory of

Dan Cunningham

&

Kathleen Harney

ENTERPRISE IRELAND

On behalf of Enterprise Ireland, I would like to welcome this important book. We were pleased to provide support for its publication because we felt that the issue of the strategic management of technology transfer and particularly the commercialisation of technology arising from research are critical both to third-level colleges and the economy itself.

Over recent years, the State has committed very significant investment into third-level research. This investment will certainly have a major impact on the development of the skill-base needed to drive the economy forward. It also has the potential to make a very direct economic impact through the commercialisation of research results that have real market potential, a key element in our *Strategy: Transforming Irish Industry 2005-2007*. In order to maximise the economic benefit from this investment, it is of critical importance that colleges have the necessary resources and expertise dedicated to the technology transfer (TT) process, commensurate with this research investment.

This book is part of the process of bringing a new level of expertise and professionalism to bear to the technology transfer process and I look forward to the debate it will engender.

Feargal O Morain
Enterprise Ireland

FOREWORD

Can Ireland sustain the momentum of its remarkable and much admired economic transformation into the next decade of the 21st century? Or are the foundations of this success too narrow and precarious to respond effectively to changing patterns of global competition?

These are the questions being asked, not only by policy-makers, but also increasingly by the higher education sector, as Ireland embarks on its next transition – from a low-cost magnet for foreign investment in Europe's evolving single market to a more self-confident and diversified knowledge-based economy, where competitive advantage has its source in research and innovation. However, they are also the very questions being asked by other countries seeking to emulate Ireland's success, including the new EU states from Estonia to the Slovak Republic, and other rapidly-developing regional economies, such as Bangalore and Shanghai. These economies, with their highly-skilled workforces, will not be satisfied for long with offshore manufacturing and the back-office operations of Western firms.

On the face of it, Ireland's growth performance appears robust. Having survived the international downturn in far better shape than most other economies, Ireland is once again posting high growth rates in productivity and employment – a simultaneous feat matched in the Eurozone only by Finland. A major factor in this performance for both countries is the comparatively high proportion of R&D-intensive exports, which, in Ireland's case, amounts to more than four times the OECD average. But this is where the similarity ends. Whereas much of the R&D content in Finland's exports is locally sourced, generating 'spill-over' effects across its national system of innovation, Ireland's R&D-intensive exports embody technology largely originating in the home-base of the multinational companies that have sited operations here to access European and global markets.

No one would doubt that the scale and quality of foreign direct investment (FDI) attracted to Ireland over the last decade is a huge achievement for a small, historically-marginalised economy, with Singapore possibly the closest modern parallel. This investment, along with European Union (EU) infrastructure support, has played a key role in driving Ireland's growth. The problem, according to recent

OECD data on the 'technology balance of payments', is that Ireland has become a *technology-taker* rather than a *technology-maker*. Further evidence is provided by Ireland's poor overall R&D performance, although it may be acknowledged that performance varies considerably across firms and sectors. Compared with the EU's ambitious but not unrealistic 'Lisbon target', which called for an R&D spend of 3% of Gross Domestic Product (GDP) by 2010, Ireland could manage only 1.12% by 2003. The EU15 as a whole did little better with 1.99%, while the US achieved 2.76% and Finland managed a formidable 3.51%.

It may be argued that R&D expenditure is no more than a proxy for performance, as it may be well-focused or badly-focused but, on all the measures we have for both the FDI and indigenous sectors, Ireland's current performance falls well short of its ambition to create a knowledge-based economy. To its credit, the Government has begun the process of building Ireland's research and innovation capacity, with the Programme for Research in Third-Level Institutions, Science Foundation Ireland, Irish Research Council for the Humanities & Social Sciences, Irish Research Council for Science, Engineering & Technology, Health Research Board, Enterprise Ireland funding schemes and, most recently, the appointment of a Chief Science Adviser. These funding bodies and programmes are beginning to make a difference, in conjunction with more competitive participation by Irish universities in EU framework programmes.

However, some major challenges remain. First, as recommended in the OECD's review of Irish higher education, not only must public research funding increase substantially in Ireland, but greater policy coherence is required across Government departments and agencies concerned with research funding and performance. This implies setting clear, consistent objectives and related delivery mechanisms in close co-operation with both the universities and institutes of technology. Second, again following the OECD recommendations, action must be taken as part of a long-term strategy to address the serious infrastructure deficit in Ireland's higher education institutions. While it makes sense for Irish universities to specialise and collaborate in key areas to achieve critical mass, there is still a resources gap to be bridged.

Finally, even with increased funding for research and infrastructure, there is a further overriding challenge for both Government and the higher education sector – effective intellectual property protection and the development of world best practice in technology transfer and commercialisation, to link research to the market. This is the subject of this path-breaking new book by Dr James Cunningham and Brian Harney of the Centre for Innovation &

Structural Change at the NUI Galway. James and Brian start from the premise that:

> The production of primary research information is not the end but the beginning of a process that continues until the usefulness of that information is realised. The commercialisation of research and knowledge for Ireland's economic benefit through effective intellectual property management and technology transfer needs to be a priority for all higher education and public research institutes and it is essential that institutes establish strong capabilities in this regard.[2]

The most important contribution of the book, however, is to examine the 'varieties of excellence' pursued by research and educational institutions around the world, and to identify those with potential application in the Irish context. The authors recognise that there is no one model to suit all circumstances, but that institutions must combine a number of essential elements in the 'third stream' activities of technology transfer and commercialisation to achieve success, whether the primary objective is to generate revenue for the institution, to build new layers of research capacity or, more broadly, to contribute to regional and national economic development.

Policy-makers in Ireland are in no doubt as to the significance of these activities:

> The exploitation of knowledge and commercialisation of research must become embedded in the culture and infrastructure of the higher education system. This requires ... new campus company start-ups, a pro-innovation culture of intellectual property protection and exploitation ... and greater links between higher education institutions and private enterprise.[3]

The question is what is the best way to achieve these new but increasingly important objectives. James and Brian go a long way towards providing practical and conceptually-rigorous answers, including investigation of the role that might be played by a 'shared services' facility, to pool resources across Ireland's relatively small-scale universities and institutes of technology.

This is a book that should be read by all those interested in the next step in Ireland's 'Celtic Tiger' experience. If future growth and competitiveness depend on the development of national innovation

---

[2]   Forfás (2004).

[3]   Enterprise Strategy Group (2004).

capacity, then the key challenge for Ireland is to link world-class research performance, not just to the global markets of today, but also of tomorrow.

Professor Roy Green
Dean Macquarie Graduate School of Management
Macquarie University, Australia
and
former Dean of Commerce
National University of Ireland, Galway.

PREFACE

A recent special advertisement supplement on Ireland in *Fortune* magazine depicted Ireland's progression from 'chronic dependency to technological leadership' (Healy & Buckley, 2005). Ireland, it went on to argue, 'now sets the pace in economic growth for EU newcomers'. In the same week, Friedman (2005) wrote in the *New York Times*: 'Germany and France will have to face reality: either they become Ireland or they become museums. That is their real choice over the next few years, it's either the leprechaun way or the Louvre'. Thus, for both the traditional economic engines of Europe, and the agile new EU accession states, Ireland's success story is said to be the model to emulate. Yet, while this attention reflects the dramatic transformation in Ireland's economic fortunes, one wonders whether, if Ireland was to ask the mirror on the wall, it would see the reflection it anticipated.

Ireland's traditional basis of competitive advantage has come under increasing threat from low-cost economies. In order to sustain its success, Ireland must turn to the creation and exploitation of knowledge as the major development path of the future. Progress along this path, however, is not unproblematic, as Ireland faces a number of key deficiencies and challenges. Delving deeper than growth in GDP to consider a few key statistics is revealing. In terms of higher education expenditure on R&D in 2000, Ireland ranked 22 out of 26 OECD countries. Ireland's business expenditure on R&D (BERD) is only 85% of the EU average and 66% of the OECD average. Critically, Ireland falls below the EU-15 average on a number of indicators of the science base said to be useful proxies for levels of innovation (for example, scientific publications, patent applications and researchers per '000 labour force). Fortunately, however, policy makers have not remained ignorant of these issues. The National Development Plan, for example, has allocated approximately €1.1 billion to research, technological development and innovation programs. What is critical is that Ireland develops its innovative capabilities and leverages the commercial potential of its research base. The key issue for Ireland's future competitiveness is the ability to make the rhetoric of the knowledge-based economy a reality for higher education institutions (HEIs) and for business.

These developments set the broad context for the genesis of this book, which was a report commissioned by the Technology Transfer Review Committee at NUI Galway to examine activities, role and mechanisms for technology transfer. The committee believed that the key question facing Irish HEIs was not *whether* they should improve, develop and maximize the effectiveness of HEI-industry collaboration, but rather *how?* This set the basis for our research, which we pursued by means of an extensive literature review, an empirical review of best practice and a series of interviews with policy-makers and university administrators. From our experiences researching this book, and in our involvement in other related projects, it is evident that there continues to be a vibrancy, enthusiasm and energy among researchers to continue to push the research frontier, often unheralded in society. Moreover, some of the funded projects currently underway in Irish HEIs have real potential to contribute to Ireland's aspiration to become a leading global knowledge-based economy. This aspiration is attainable, but we have a long way to go if we are to be taken seriously as a knowledge-based economy. Significantly, we would argue that, for HEIs to contribute to this attainable aspiration, they have to engage and mobilise their own institutions, while simultaneously engaging and informing society about their core missions and the value of pursing third-stream activities and disseminating research for the public good.

This book is designed to highlight the issues that are involved in strategically managing technology transfer. In so doing, we appreciate that there is no 'silver bullet' when it comes to developing and exploiting third-stream activities. Instead, we draw on what we term the 'varieties of excellence' evident in best practice institutions. The book explores the structures, activities and mechanisms for technology transfer, while also reviewing conditioning factors that shape the parameters of technology transfer initiatives. These include stimulants and barriers to technology transfer, as well as the difficulties in evaluating technology transfer outcomes. These efforts culminate in the final chapter, which presents a framework to guide the strategic management of technology transfer, ensuring that a necessary balance is reached in third-level institutions between commercialisation activities and more traditional activities of teaching and research dissemination.

Overall, we contend that the creation, exploitation and commercialisation of research are critical if Ireland is to overcome competitive threats and sustain its economic momentum. The tendency to highlight the Irish model as a benchmark for emulation may be somewhat premature. If Ireland really is to be an exemplar of excellence for the rest of Europe, it must believe in, and realise, the words of Juan Enriquez of Harvard University, who wrote at the start of this decade: 'the future belongs to small populations who build empires of the mind'. Ultimately, we hope this text goes someway towards facilitating this objective. At a minimum, it should set the foundation for debate and raise awareness in an area central to maintaining Ireland's successful economic trajectory.

We hope that you enjoy reading this book and that it enables to you reflect on your experiences and those of your institution. We welcome any comments, queries or suggestions, which you can e-mail to james.cunningham@nuigalway.ie or b.harney@jbs.cam.ac.uk.

James Cunningham
Brian Harney

Centre of Innovation and Structural Change (CISC)
and
Department of Management
National University of Ireland, Galway

January 2006

1
TECHNOLOGY TRANSFER IN CONTEXT

*Ireland's economy has performed exceptionally well over the
past decade and we have built a world-class reputation across a
range of technologies. The challenge now is to maintain and
grow this performance. Our future depends on how well we
manage the transition to a world where knowledge and ideas are
more important than bricks and mortar. Every economic region
is trying to capitalise on new technologies and new markets. If
we are to succeed, we have to do the same, but we must be more
creative and flexible than our competitors.*
Mary Harney, Tánaiste, July 2004[4]

INTRODUCTION

Increasingly, it is acknowledged that universities and their research
activities may form the perfect catalyst in propelling countries along the
trajectory towards becoming knowledge-based economies. It is
predicted that those nations with strong research systems and the
capacity to leverage the commercial opportunities of their research will
be best placed to prosper both economically and socially. These are
arguments that carry particular force in an Irish context, where public
policy has recently focused on enabling criteria to facilitate this
development. In particular, much emphasis has been placed on
innovation, which has become the defining challenge for global
competitiveness (Porter & Stern, 2001). The National Development Plan
(2000: 128) argued that: 'the government accepts that there is a strong
link between investment in the research and innovation base of the
economy and sustained economic growth'. Such an emphasis provides
universities with a 'strategic role' in the context of knowledge-based
economies, given their potential to upgrade the skills and knowledge of
the labour force and their contribution to producing and processing
innovation through technology transfer (Hernes & Martin, 2000: 14). The
development of innovative capabilities in Ireland, however, still remains

---

[4]   Mary Harney at the launch of the findings of Enterprise Strategy Group, quoted
     in *Irish Independent*, Digital Ireland Supplement, 9 July 2004.

a challenge, as noted by the *Global Competitiveness Report* (World Economic Forum, 2001): 'Ireland, which has been tremendously successful in attracting foreign investment for manufacturing, now faces the need to justify higher wages and higher local costs without yet having developed a world-class innovation structure'.

Higher education institutions (HEIs)[5] can be powerful motors for the technological and economical development of industrial branches and regions (Shattock, 2001). This, coupled with the fact that industry is becoming more science-based and knowledge-driven, means that 'collaboration between fundamental researchers in universities and applied researchers in industry is the engine of modern innovativeness and competitiveness' (EIMS, 1995).

HEIs in the Knowledge–based Economy

Ireland's future economic development will be strongly influenced by 'the increasing role of knowledge as a driver of economic development and an influencer of new products' (Enterprise Strategy Group, 2004: xi). The Enterprise Strategy Group (ESG) presented a model of the elements that would enable sustainable development in Ireland in the context of a knowledge-based economy (**Figure 1.1**). Inherent within this model is the multiple roles of universities, in terms of enabling the essential conditions for success (particularly innovation and entrepreneurship and management capability) and in creating and maintaining the sources of competitive advantage (expertise in technological product and service developments and world-class skills and training).

Clearly, HEIs have the potential to be a significant source of knowledge and capabilities within the knowledge economy (Molas-Gollart *et al.*, 2002). In recognition of this fact, one of the five sources of competitive advantage identified by the ESG was 'expertise in technology: product and service development' (detailed in **Table 1.1**). The establishment of linkages between industry and science are considered paramount to the realisation of an economy that emphasises the role of knowledge and technology in driving productivity and economic growth (Beesley, 2003: 1522). A new organisation field broader than the traditional organisational field consisting of industry, government and research has emerged (Leydesdorff & Etzkowitz, 1996). This is best captured in the concept of a triple helix system, which recognises that the future location of research and technology transfer reside in a triple helix of university-

---

5   The term 'HEI', as used throughout this book, denotes Universities, Institutes of Technology and Institutes of Further and Higher Education.

industry-government relations (Leydesdorff & Etzkowitz, 1996; Etzkowitz & Leydesdorff 1999). This model acknowledges that overall innovation performance of the economy is not dependent on how specific institutions perform but rather, overall performance depends on how they interact with each other as elements of a collective system of knowledge creation and use (OECD, 1994).

Figure 1.1: Sources of Competitive Advantage in the Irish Knowledge-based Economy

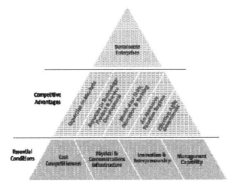

Source: Enterprise Strategy Group (2004).

Table 1.1: Expertise in Technology: Product & Service Development

Competitive Advantage	Characteristics
Expertise in technology: Product & service development	Development of world-class capability in focused areas of technology and in innovative techniques to drive the development of sophisticated, high value products and services.

Source: Enterprise Strategy Group (2004: xv).

THE TRIPLE HELIX MODEL[6]

The triple helix model is based upon a spiral model of innovation, which allows for the three institutional domains to interact and share mutual relationships with all three spheres. The triple helix, however, is 'more complex than the mutual interactions between the "double helices" upon which it rests' (Leydesdorff, 2000). The triple helix is

---

[6]   The authors wish to acknowledge Will Geoghegan, a Research Fellow at CISC, NUI Galway, for his contribution to this section.

comprised of various linkages that take place at numerous parts of the innovation process and is categorised by this 'spiralling' of the three spheres, which are seen to be increasingly involved in each others' activities. As different innovation systems theories have evolved, and the linear models have been refuted, the triple helix has risen as an analytical tool to assess the interaction between university, industry and government (Godin, 2005).

The concept of the triple helix emerged from a failure to concisely locate and define the processes of innovation. The triple helix thesis is said to be organised around three intertwined dynamics, namely:

♦ Institutional transformations.

♦ Evolutionary mechanisms.

♦ The new position of the university.

Critically, the triple helix model affords central importance to the role of the universities in a context where the focus has previously been exclusively at the firm level of analysis.

Etzkowitz & Leydesdorff (2000) note that 'typifications in terms of 'national systems of innovation' (Lundvall, 1998; Nelson, 1993); 'research systems in transition' (Ziman, 1994, Cozzens *et al.*, 1990), Mode 2 (Gibbons *et al.*, 1994) or 'the post-modern research system' (Rip & VanderMeulen, 1996) are indicative of flux, reorganization, and the enhanced role of knowledge in the economy and society. Specifically, Etzkowitz (2002) cites four reasons for the paradigm shift facilitating the emergence of the triple helix dynamic:

♦ **Internal transformation of each of the helices:** An example of this is the emerging role that universities are beginning to play in a wider societal context (departing from previously only training and educating students to now beginning to apply the knowledge and research to commercial avenues, therein transcending traditional boundaries of university-industry linkages). Another often cited example involves companies initiating lateral ties through strategic alliances. In terms of government, an example is the increasing propensity of governments to offer venture capital to firms.

♦ **The influence of one helix upon another:** An example of how one part of the helix may seek to influence another part is the Bayh-Dole Act of 1980. This was a US governmental industrial policy providing universities with ownership of federally-funded research, thereby enabling them to retain any royalties obtained from commercialisation activities.

♦ **The creation of a new overlay of tri-lateral networks and organisations from the interactions among the three helices:** Etzkowitz (2002) cites how this dimension is especially important at the level of regional industrial clusters. Previously, these may not have been initiated or obtained interest or investment.

♦ **The recursive effect:** The recursiveness of triple helix networks effect the spirals from which they emerge and society in general (Etzkowitz, 2002). For example, the capitalisation of knowledge may change the way in which the researcher or scientist views the results of their work and thus the role of the university is transformed with regard to the other two spheres.

The triple helix is an analytic tool that adds to the description of the variety of institutional arrangements and policy models an explanation of their dynamics. It is this dynamism that leads to an emerging overlay of communications, networks and organisations among the helices (Etzkowitz & Leydesdorff, 2000). Piekarski & Torkomian (2005) underline that it is these 'dynamic relations' that represent the true nature of innovation systems.

Roles of Industry, Government & Universities in the Triple Helix

The Role of Industry within the Triple Helix

The specific role of industry and firms is to 'pressure government into provision of an environment, which will help maintain the firm's competitive position' (Nedeva *et al.*, 1999). In the past, industry was considered the sole exploiter and generator of innovations. This dynamic is changing however, with the increased activity of the 'entrepreneurial university' in the innovation process (Melera & Arcelus, 2005).

Nedeva *et al.* (1999) researched whether industry saw itself as a beneficiary or as a benefactor when it came to supporting university research. They found that there is no simple answer and that a single firm employing the university may garnish far more than a firm funding basic research that did not employ specific deliverables or criteria. Porter (1990) lends support to the idea that the role of industry is vital. In his book on the competitive advantage of nations, Porter (1990) advocates that four factors – firm strategy, structure and rivalry; factor conditions; demand conditions; and related and supporting industries – help to explain where a nation may lie with regard to its competitive positioning *vis-à-vis* other nations. However, he does not give universities central importance as a determinant in sustaining a comparative advantage (but sees them as an important element within factor conditions) and sees the role of government as indirectly influencing all of the dimensions mentioned above.

The Role of Government within the Triple Helix

The governmental role in fostering innovation can be seen as pivotal. Mowery *et al.* (2001) cite the often-neglected enormous post-war US federal investment in university research as critical in building the US' foundation for success in commercialisation activities. According to Piekarski & Torkomian (2005), governments must allow for the smooth functioning of markets and for the externalities associated with R&D investments, act as an important agent in some areas of economy, and also remove systemic imperfections in their innovation systems. Further, they note that the social return on public research funding is often greater than the pure private rate, using Den Hertog & Roelandt (1999) to substantiate this assertion. According to Etzkowitz (2002), the role of government is expanding, not only in relation to macro factors, but increasingly to encompass the micro conditions of innovation.

The Role of Academia within the Triple Helix

Technology transfer and the commercialisation activities of universities are vital to the innovation process. Etzkowitz & Leydesdorff (1997) begin their book *Universities & the Global Knowledge Economy: A Triple Helix of University-Industry-Government Relations* as follows: 'The development of academic research capacities carries within itself the seeds of future economic and social development in the form of human capital, tacit knowledge and intellectual property'. They argue that the responsibility of channelling knowledge flows has become the task of the university, therein changing the contemporary university structure and function. Etzkowitz (2003) explicitly informs us that we have transcended from the 'research university' to the 'entrepreneurial university'. He states that, in its 'embryonic' form, the entrepreneurial university surfaced in the late 19th century, when a lack of a formal research funding system forced academics into vying for resources to support their research. Etzkowitz (2003) sees the evolution in the US and Europe contrastingly, emphasising a 'bottom-up' approach in the US, as opposed to a 'top-down' approach in Europe. He goes on to suggest that the so-called 'national champion' view of economic success is beginning to rescind, opening up the way for a thinking that evolves around clusters of closely associated firms, which usually is associated quite heavily with a focal university or other research institute. Powell *et al.* (1996) underlines that there is an increasing trend in which new and established companies are beginning to foster and enhance relationships with educational entities. Similarly, Godin & Gingras (2000) refute Gibbons *et al.*'s (1994) claim that the university is receding, arguing instead that universities have been able to stay at the centre of the knowledge production system through exploiting collaboration mechanisms. Typically, these relationships stem from universities establishing a tie with industry toward technology transfer by means of consulting, research by contract, and patent license (to existing firms), and spin-offs (stimulating the creation of new firms) (Piekarski & Torkomian, 2005).

Overall, the triple helix model positions HEIs and their activities as critical in facilitating the system of knowledge production in the wider innovation system. As noted by Etzkowitz & Leydesdorff (2000) 'the triple helix states that the university can play an enhanced role in knowledge-based economies'. Innovation, therefore, cannot be seen as the product of one institutional sphere but rather is the product of a 'system of innovation' (Leydesdorff & Etzkowitz, 2001). In this respect, the different helices are seen to be undergoing a common change in direction, which will stimulate both competition and cooperation. To understand these developments better, it is necessary to consider factors conditioning this change, which include the increased impetus for research commercialisation, business demands for research excellence as well as the benefits and concerns that have emerged as a result.

THE IMPETUS FOR INCREASED EMPHASIS ON RESEARCH COMMERCIALISATION

The attention directed at the potential economic and social value that maybe leveraged from HEI research has increased dramatically in recent times. The intertwined forces of globalisation and technological developments have altered the structure of the industrial landscape. Some of the major factors conditioning these changes include:

♦ **Co-opetition and collaboration:** There is an increased emphasis on co-opetition and collaboration. It is no longer possible for any organisation, regardless of its size, to rely solely on internal competence. IBM, for example, is part of many co-operative efforts in product and technology development (Lundvall, 2002: 3). Multiple opportunities are arising as a result of the way companies are undertaking research and development. Carayannis & Alexander (1999) argue that the future secret of business survival and prosperity lies in strategic partnering and co-opeting successfully rather than in outright competition. University-industry relationships are a central a part of this wider process of collaboration (Smith, 1991).

♦ **Developments in ICT:** The development of ICT or the 'death of distance' has meant that economies have become much more open. This has sparked what has been referred to as 'the globalisation of knowledge' (Mason, 2003). Consequently, companies are operating well beyond their original home base and are conducting R&D activities at multiple locations across the globe.

The interaction between HEIs and business has been a classic research and policy theme among OECD and other international organisations (Lundvall, 2002: 9). More recently, considerable attention has been focused on research commercialisation aspects of this relationship. This focus on commercialisation can be attributed to the following:

◆ **Innovation and multiple actors:** In the past, innovation was depicted as a straightforward conversion – investment in basic science to economic growth, passing through applied science, technological developments and marketing (Lundvall, 2002). Now there is now an appreciation that innovation is open and characterised by complex feedback loops, uncertainty and pure serendipity (Etzkowitz & Leydesdorff, 1999; Fassin, 2000; Graff *et al.*, 2001; Kline & Rosenberg, 1986). Furthermore, innovation is now generally accepted as being the result of the input of multiple actors (Den Hertog, 2003: 95).

◆ **Bayh-Dole Act 1980:** This legislation allows US universities to patent inventions funded from federal money and to retain the royalties that these patents generate (Coupe, 2003). It has been cited as the catalyst that has encouraged commercialisation activities within US universities. Universities can now establish ownership rights and generate a new source of income from their research activities. In the past, universities had very few incentives to undertake commercialisation activities based on their research results. Between 1980 and 2003, more than 50 universities have generated $2+ million in licensing revenues, and a handful have generated revenues in excess of $100 million annually (AUTM, 2003; Brint, 2005).

◆ **Scientific breakthroughs:** Developments in areas such as biotechnology and related life science fields have led to a dramatic shortening of the time from scientific breakthrough to commercial use. Similar, but less dramatic, changes have taken place in the fields of software and communication technologies (Lundvall, 2002: 9). Consequently, the interface between basic science and product development increasingly includes aspects of the social sciences (Etzkowitz & Leydesdorff, 1999). The OECD (2003: 55) has noted that 'the explosion of patenting, and especially of PRO patenting, is often attributed to the rise of biotechnology and other medical sciences as commercially relevant fields'.

◆ **Competitive threats:** Internationally, and particularly in an Irish context, the competitive threat from Asia and the new EU accession states has added to the sense of urgency to focus on the commercialisation of research activities. Indeed, it is threats from these economies that are said to have focused government attention on differentiating Ireland from our EU brethren through knowledge as a key competitive tool (Cope, 2004: 12). Capabilities in patenting and licensing are therefore critical as potential sources of future competitive advantages (Connelan, 2004).

Clearly, if fundamental research is the seed of innovative activity, then HEIs are the seed banks and play an essential role in sustaining an innovation led economy (Sheehan & Wyckoff, 2003: 32). These changes have resulted in an increased intensity of university-industry linkages (Hernes & Martin, 2000). The changing role of HEIs, while having multiple sources, has been strongly influenced by the blurring distinction between basic and applied research and the fact that patterns for the conduct of R&D in science and technology have widely changed.[7] These factors, in turn, have directed the attention of businesses to HEIs as sources of knowledge and innovation and as potential partners to facilitate them in their quest for competitive advantage.

BUSINESS DEMAND FOR RESEARCH EXCELLENCE

Business needs to access relevant research in order to build expertise and capability in product and service development (ESG, 2004: 68). The requirement for research excellence traditionally has been a necessary, if not sufficient, condition for business success. In the light of intensified competition and shortened product life cycles, however, the ability of business to provide this crucial input to innovation has been challenged. This has led to a number of changes and demands in the nature of research and development activities, namely:

♦ **Innovation:** Companies must be able to innovate, create and commercialise a stream of new products and processes that shift the technology frontier, progressing as least as fast as their rivals (Porter & Stern, 2001). Successful innovation leads to new products, new markets and to commercial and financial success (Fassin, 2000).

♦ **Intellectual Property:** IP in various forms (patents, trade secrets, or information) has become increasingly important as a competitive force (ESG, 2004; Nelson, 2001). Thus, in the knowledge economy, the value of goods and services is not only measured by their price, but also by the amount of new knowledge they embody (Hernes & Martin, 2000; Mason, 2003).

---

[7] Basic research is said to have stronger public goods properties, historically seen to have been the province of universities while industry is left to concentrate where its interests and expertise lie, in applied research and development (Graff *et al.*, 2001). While reference is made to Mode 1 and 2 research, this dichotomy is used heuristically to define basic and applied research, although the two kinds of research are rarely so distinct from one another and in fact the conventional distinctions are blurring (see Rosenberg & Nelson, 1994).

♦ **Outsourcing:** Companies have begun to recognise the value of outsourcing more of their upstream research to university laboratories. The rationale provided for this is related to traditional economic arguments in terms of the uncertainty and inappropriability of such basic research (Allan, 2001).

The impacts of these changes are evidenced by the relative increase in the provision of funding from private business to R&D in HEIs (Hernes & Martin, 2000). For many firms, building in-house R&D capacity is a challenge. They lack resources, not only to conduct R&D, but also to absorb new developments coming from outside.

The European & Irish Experience

According to European Commission figures, gross expenditure on R&D (GERD) expressed as a share of GDP (GERD/GDP) in 2001 was highest in Japan, followed by the USA, with Europe lagging significantly behind (see **Figure 1.2**). The largest component of the gap between the USA and the EU-15 can be attributed to business funding, but public financing is also significantly less in Europe than in the USA. Industrial investment in R&D is unlikely to grow significantly while the EU investment in high-quality frontier research lags behind in fast-growing areas where science and technology are closely intermingled (Forfás, 2004).

Figure 1.2: R&D Expenditure as a Proportion of GDP, by Source

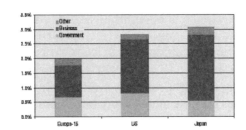

Source: Towards a European Research Area – Science, Technology & Innovation – Key Figures 2003-2004, European Commission, 2004.

In an Irish context, BERD increased to €1,076 million in 2003, up from €901 million in 2001, an annual increase of 9.2% between 2001 and 2003 compared with 7% between 1999 and 2001 (Forfás, 2005). Five sectors accounted for nearly 80% of expenditure on R&D in 2001 (see **Figure 1.3**). A recent report to the Inter-departmental Committee on Science,

Technology & Innovation (Forfás, 2004) stated that, for Ireland to sustain its rate of development, BERD will have to increase to €2.5 billion by 2010, equivalent to 1.7 % of GNP.

Figure 1.3: R&D Expenditure by Sector, 2001

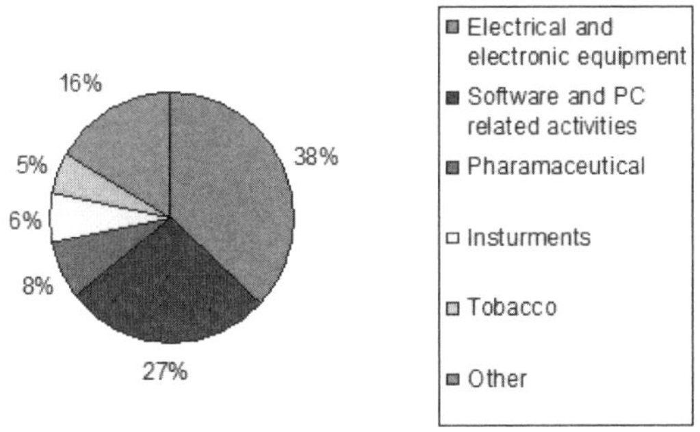

Source: Adapted from Forfás (2003).

R&D in the higher education sector reached €238 million in 2000, up from €204 million in 1998, with universities accounting for some 80% of higher education-based R&D. In 2002, expenditure increased by 35% to €322 million. While the 2004 figures are not yet available, projections indicate that expenditure is expected to reach €500 million (Forfás, *Annual Report*, 2004). In comparison to Europe, however, investment in R&D and innovation in Ireland is relatively low. At 0.88% of GDP, Ireland's BERD is 73% of the EU average and is 57% of the OECD average (ESG, 2004). In terms of higher education expenditure on R&D, in 2000 Ireland ranked 22 out of 26 OECD countries (ICT Ireland, 2004a). Most recent figures suggest more encouraging signs, as the share of total expenditure invested in basic research by business rose to 8.9% in 2003, increasing from the 4.4% share recorded in 2001. A worrying fact, however, is that Ireland has low R&D intensity in sectors that are of particular importance to the Irish economy (see **Table 1.2**).

Table 1.2: BERD as a Percentage of the Output by Sector, 2001

	Ireland	OECD Average
Electrical & Electronic Equipment	1.4	5.6
Pharmaceuticals	1.3	11.5
Instruments	1.3	7

Source: Forfás (2003).

Benefits to Business of Collaborating with HEIs

Chen (2005) contends that 'direct relationships between university and industry can bring important competitive benefits to firms'. HEIs can form a perfect bridge between technology and companies (Porter & Stern, 2001). The potential ways in which businesses can gain competitive advantage from working with universities include:

♦ Increased access to new university research and discoveries. The best academic researchers are usually international in their scope and range of knowledge. At a local level, researchers also have much experience and established networks (Lee, 2000).

♦ Multidisciplinary access to facilitate achievement of excellence.

♦ The ability to leverage the research dollar, particularly in the case of more basic research, where the early stage is more uncertain, and there may be along time lag before any return on investment (Graff et al., 2001). Indeed, many high-tech commercial successes come from developments that are impossible to foresee with long time horizons (Gibbons et al., 1994).

♦ Shared R&D expenditure, since the largest part of R&D development expenditure (*circa* 50%) is paid in wages (Sheehan & Wyckoff, 2003).

♦ Access to expertise in terms of human resources. Universities appear to be an propitious site for innovation, due to such basic features as their high rate of flow-through of human capital in the form of students, who are a source of potential inventors (Etzkowitz, 2003). Collaboration provides an opportunity to spot and recruit the brightest young talent. Massachusetts Institute of Technology (MIT) believes that one of the motives of those corporations that invest in its industrial liaison program is the knowledge that this gives them knowledge about the best students who are passing through the system.

♦ Access to specialised consultancy.

♦ Access to established networks and communities of knowledge exchange.

♦ Access to capabilities where organisations lack the expertise required for the innovation process (Georghiou, 2004).

♦ Enhanced reputation from working with leading edge researchers and HEIs.

In the last quarter of a century, transforming research into new goods and services has become more pronounced. 'For existing firms, the university can be both a problem and a solution, a technological competitor and a technological saviour; and the entire difference turns on relationships formed and the intellectual property rights one obtains in an arcane sounding process known as technology transfer' (Graff *et al.*, 2001: 3). The state of play is similar to that depicted by Chesborough (2003) who argued that 'increasingly, the university system will be the locus of fundamental discoveries and industry will need to work with universities to transfer these discoveries into innovative products, commercialised through appropriate business models'. Yet, as noted by the ESG (2004: 93), 'to ensure a steady flow of ideas, mechanisms that foster linkages and collaboration between academia and enterprise are critical'. Reflecting this, there has been a considerable literature about university-industry relations making reference to the triple helix, Mode 2 interdisciplinary research and the concept of 'entrepreneurial science' (Gibbons *et al.*, 1994). This has led to a fundamental re-conceptualisation of the role of the university in modern societies, as discussed in the following section.

THE GROWTH OF THIRD-STREAM ACTIVITIES: BENEFITS & CONCERNS

The perception of HEIs as merely institutions of higher learning is gradually giving way to the view that universities are important engines of economic growth and development (Chrisman *et al.*, 1995). While, in the past, HEIs have focused on two streams of activities – research and teaching – 'third-stream' activities now form an important part in terms of universities' missions and objectives. University technological innovation and technology transfer are seen to be key contributors to the new wave of economic developments (Fassin, 2000: 31). In the UK, Lambert (2003) has termed this trend as one of 'casting off their ivory tower image'.

Benefits

HEIs are playing a more active role in the process of technological innovation by licensing inventions and discoveries to industry (Fassin, 2000). This intensity in terms of collaboration activities is further enhanced by pressures faced by universities, such as the increased costs of scientific equipment and insufficient government funding. Such pressures have forced HEIs to look to a variety of sources for new means of financing research. Furthermore, an enhanced expectation of economic payoff, due to R&D results, has influenced the attitudes of academic staff and institutional managers, leading to increased collaboration with industrial partners, especially in fields such as biotechnology, medicine and software development (Hernes & Martin, 2000).

The perceived benefits of linkages with business for HEIs include:

♦ Exploitation of research results. Research results that otherwise might remain unexploited, are disseminated to the broader society.

♦ Creation of spin-offs can help generate employment.

♦ Commercialisation activities can provide an additional source of revenue for the university. This is especially important as universities have become more dependent on external sources of research funding (Kauonen & Nieminen, 1999).

♦ Greater cross-fertilisation between entrepreneurial faculty and industry, which in turn can improve teaching, broaden research horizons and facilitate in developing an up-to-date curriculum.

♦ Linkages may serve to enhance an institution's reputation. Closer interaction with the private sector may also improve the quality of research, enable a world-class research ranking, and the development of strong linkages with companies and funding agencies (Allan, 2001; OECD, 2003; Plonski, 2000; Siegel *et al.*, 2003).

♦ This enhanced reputation and/or linkages with high profile companies may facilitate the university in attracting the best students (Fassin, 2000: 40; Schindel, 2002).

♦ As well as intrinsic benefits, such as profile enhancement, researchers may obtain financial rewards from licensing agreements, selling IP or even be headhunted by industry.

♦ Department/research group benefits include generation of research income, which can facilitate more research support, access to external resources, specific consultancy projects, development of a critical mass by research discipline and a good international profile, which may assist in attracting staff (Lee, 2000; Pandya & Cunningham, 2000).

♦ Collaboration with business may also enhance teaching content. Close contact with industry problems is essential for the vitality of research and teaching. Engaging with the world of practice, therefore, keeps teaching relevant and current.

Concerns

Yet, if the opportunities for, and benefits from, conducting HEI-industry research collaborations have soared, so have concerns about their implications. The rising number, pervasiveness, variety, and importance of these partnerships have heightened their impact, while raising the stakes involved. As universities pursue additional funding sources and companies seek continued competitive advantage – and as both try to keep up with the accelerating pace of change – these partnerships have become an increasingly critical means toward achieving key objectives. This complexity raises a number of policy issues concerning the potential unintended effects of such activities – for example, diversion of energy and commitment of staff, impact on university missions and potential distortion of research to more lucrative fields. Indeed, the OECD (2003) has noted a substantial backlash in those countries that have not given these issues due consideration.

STRATEGICALLY MANAGING THIRD-STREAM ACTIVITIES: AN ISSUE OF BALANCE

As knowledge-producing institutions, HEIs play a critical role in the emerging 'third industrial revolution', based on information and genetics technologies (Gulbrandsen & Etzkowitz, 1999). Some have coined the term the 'entrepreneurial university' (Clarke, 1998) or made reference to the educational-industrial complex (Graff *et al.*, 2001). Yet, while the commercialisation of property owned by HEIs is an important component of third-stream activities, it is only one among the many functions that link HEIs with society (Molas-Gollart *et al.*, 2002). Commercialisation activities should be promoted as *complementing*, as opposed to *cannibalising*, existing activities. Essentially, the issue is one of *balance*. The emphasis should be on balancing the objectives of managing intellectual property rights, developing new revenue sources, and accommodating faculty interests, while simultaneously maintaining norms related to the conduct of academic research and the dissemination of research findings (Feldman *et al.*, 2002; OECD, 2003: 104). In acknowledgement of this, Beesley's (2003: 1527) interpretation of the triple helix emphasises *interaction* rather than *intersection* between different players, noting that 'while the sectors are interacting, they are maintaining their current structures and identities, complete with norms, cultures and reward structures'. Research by Mowery & Arvids (2002), looking at the content of academic research in the wake of Bayh-Dole Act 1980, found that there was little change in terms of content but rather simply more intensified marketing efforts, particularly at the University of California and at Stanford.

HEIs, both nationally and internationally, have evolved to meet multiple objectives, including education, contributing to works of culture, nurturing arts, hosting the intellectual discourse and conscience of society, and generating new scientific knowledge for the benefit of both the economy and society at large (Graff *et al.*, 2001). If correctly understood and strategically managed, third-stream activities, particularly research commercialisation, can offer an important contribution to achieving HEIs' missions. In this role, commercialisation activities should be seen primarily as a means to generate public value, with raising funds as a second objective. Pursuit of multiple objectives and the concomitant reliance on a variety of funding sources enable HEIs to establish research units that are quite unique in their capabilities and that have distinct relative advantages in terms of capacity and cost-effectiveness. Furthermore, the various activities of HEIs should be seen as mutually reinforcing and inclusive (Graff *et al.*, 2001: 7). Co-operation refers to mutual influence: 'to be able to co-operate, the university must open its doors to the surrounding community and allow the needs and ideas, that spring from its new contacts, to become the object of critical reflection, and where relevant to be integrated into teaching' (Lindberg, 2000).

In order to locate these arguments to the Irish context, it is necessary to review the key macro trends that characterise the Irish research landscape. In doing so, we can understand better the specific implications of the changing roles of HEIs for Irish HEIs' leaders and administrators.

The Research Environment in Ireland

Despite many existing initiatives in Ireland, overall collaboration between enterprise and academia has been limited. This is particularly the case when Ireland is compared with EU member states of a similar size, where co-operation and networking are well-established practices – for example, Finland, Denmark, Austria (ESG, 2004: 68). In 2001, R&D spend in the higher education and public research sector in Ireland was €422 million or 0.4% of GNP, compared to the EU average of 0.66%. Of particular concern is the deficit of researchers per '000 labour force in Ireland. Ireland ranks around the middle of the European league with 4.87 researchers per '000 labour force, lagging substantially behind innovators such as Sweden with 9.1 researchers per '000 and Finland with 9.6 researchers per '000 (see **Figure 1.4**) (Connelan, 2004). A significant increase in the number of researchers in Ireland – from 5,796 in 2001 – is a necessary condition to drive the creation of intellectual property and patents. Recent media accounts of the low intake into science and engineering degrees, despite government efforts in this area, are not encouraging.

One approach to evaluating the benefits of publicly-funded research is to analyse the scientific publications cited in patents. In the US, this has indicated a strong and growing reliance by industry on the results from publicly-funded research. This has had a significant impact on government research policy in the US and deserves to have a similar impact on research policy in Ireland (HLEG, 2005: 28).

The Irish research system carries a legacy from years of under-resourcing, which has resulted in high levels of fragmentation, low levels of collaboration and lack of critical mass. The performance of Ireland's science base can be judged by reference to two commonly-used measures: scientific publications and patents. Ireland falls significantly below international standards on each of these measures. The contrast with smaller European countries that, like Ireland, aspire to being knowledge-based, is particularly stark (see **Table 1.3**).

Figure 1.4: Researchers per '000 Labour Force, 1999

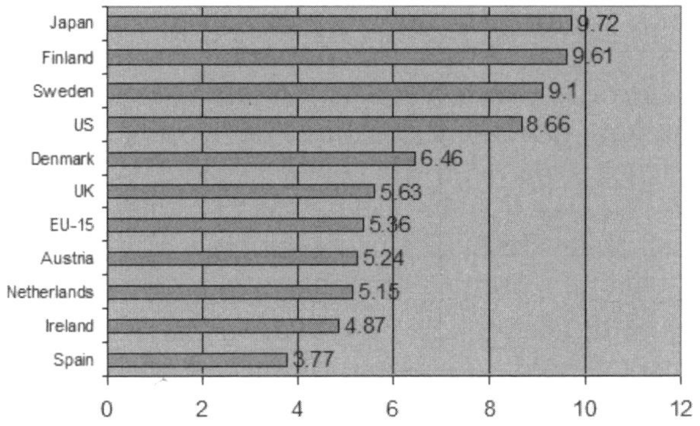

Source: Connelan, 2004.

Table 1.3 Measures of Performance of the Science Base

	IRL	EU-15	SWE	DEN	FIN	US
Scientific Publications per million population (2002)	647	673	1,598	1,332	1,309	774
European Patents applications per million population (2000)	61.6	128.4	248.2	151.3	258.6	103.6
US Patents granted per million population (2002)	32.1	71.2	187	83.7	158.4	300.5

Source: European Commission, Key Figures 2003-2004.

IMPLICATIONS FOR HEI LEADERS &

ADMINISTRATORS

Increasingly, it is acknowledged that developing and improving technology transfer and commercialisation activities is of critical importance for HEIs. Funding difficulties, coupled with recognition of the potential returns from business partnerships and spin-offs, are forcing terms such as 'technology transfer' into the lexicon of mainstream academia. The question Irish HEIs are now facing is not *whether* they should improve, develop and maximise the effectiveness of HEI-industry collaboration, but rather *how*? Concerns about how best to address issues of funding, patenting, research commercialisation, managing start-ups, developing codes of conduct, as well as the governance and leadership of technology transfer, are ones that not only face Irish HEI leaders but industry collaborators and policy-makers alike. Strategically developing third-stream activities involves negotiating a number of managerial tensions that include:

♦ **Risk *vs* reward:** The primary objective of publicly-funded HEIs is the development of public goods and the protection of their intellectual property. HEIs have to balance these goals with an increasing requirement to exploit their research output commercially. Successful universities in the 21st century will be those that continue to excel in teaching and research, while at the same time developing complimentary third-stream activities to exploit research for the public good. It is in this context that questions of co-ordination, adherence to university regulations and recognition in terms of success of institutional resources or of a financial return to the institution become issues of great sensitivity (Shattock, 2001). This has been the case, particularly in relation to company formation activities, which can be viewed as a double-edged sword: acting as potentially the most lucrative and beneficial transfer technology mechanism, while at the same time often requiring the most investment and attendant risks, thereby potentially leaving HEIs most exposed. This tension can be managed by putting in place solid governance structures and transparent policies that ensure an appropriate balance in third-stream activities

♦ **Individual research efforts *vs* funded priority areas:** At the heart of HEIs is academic freedom, particularly in relation to research efforts. Academics have the freedom to participate in national and internationally-funded projects. HEI leaders have to ensure an adequate balance between ensuring support for the commercialisation of priority-funded research, while simultaneously encouraging and supporting research that does not fall within the remit of institutional priority areas or is not in receipt on national or international research funding. HEIs can manage this tension by ensuring that academics become involved in commercialisation or in specific research areas based purely on their voluntary desire to do so. Critically, HEI leaders should also remember that technology transfer maybe more appropriate to specific fields and is only one of a range of third-stream activities conducted by universities. Moreover, there are multiple channels and methods, direct and indirect, formal and informal for technology transfer.

♦ **Scholastic endeavours** *vs* **academic entrepreneurship:** The concept of 'academic entrepreneurship' has emerged to capture the trend for individual researchers to become increasingly involved with the commercialisation of their research results, through licensing a patent or by the creation of a spin-off company. Traditionally, HEIs' missions have focused on teaching and research excellence and HEIs' promotional structures reflect this perspective. With increased emphasis on commercialising research within HEIs, and from funding agencies nationally and internationally, HEI leaders will face the task of reflecting these changes in an equitable manner, in terms of promotion systems and resource allocation. HEIs thus will have to organise themselves in a way that will allow them to exploit academic entrepreneurship in the most effective manner (Lambert, 2003: 6).

♦ **Local initiatives** *vs* **regional/national collaboration:** Currently, all HEIs in Ireland act relatively independently in managing their technology transfer activities (Forfás, 2005). The reality facing all Irish HEIs, however, is that they will not be in a position to provide all the services required for university-industry collaboration. Given their deficiencies in expertise, particularly in aspects of commercialisation and knowledge of the patent process, HEIs will be under increasing pressure to develop more co-operative regional or national mechanisms. The EU, in particular, emphasises the value of collaborative efforts in providing economies of scale and expertise. HEI leaders, therefore, will have to develop clear policies and procedures in relation to their participation and activities in a meso-level structure, while at the same time ensuring that they maintain a close relationship with researchers on the ground in their own institutions.

The nature of these challenges and tensions are best understood in the context of the broader cultural and institutional re-conceptualisation of universities, as illustrated in **Table 1.4**.

In the US, the AUTM (2002: 7) has stated that the mission of technology transfer has permeated all parts of academia: 'even the smallest colleges and universities are creating the infrastructure to translate the fruits of their research into products that serve the public good'. The paradigm shift to a more entrepreneurial university therefore appears to be a real one. This is the case not only in the US, but also in Ireland and rest of Europe where new arrangements have to be negotiated so that information can flow smoothly from a lab in the university to a company (Etzkowitz & Leydesdorff, 1999). Passive approaches of fortuitous technology transfer that serve to reinforce the 'ivory tower imagery' need to be replaced by more proactive approaches. In an Irish context, this requirement was highlighted by the Enterprise Strategy Group (2004: *xvi*): 'close to market and applied research capabilities must also be promoted, to facilitate greater synergy between those who generate knowledge and those who transform it into saleable products and services'.

Table 1.4: Re-conceptualising the University: Characteristics of Established Research Universities & New University Models

Characteristics	Established Research universities	New University Models
Agents	Individuals and small teams	Large multidisciplinary groups
Means	Agency and Foundation Grants, Fellowships	Concentration of large-scale economic resources and/or political support networks
Orientation of agents	Disciplinary, sub-disciplinary	Interdisciplinary
Underlying Dynamic	Cumulative Progress in fields of formal knowledge	Constant innovation in economy and society
Criterion of Success	Rank in national ratings	Contribution to economic and social progress
Legal frame	Tenure and promotion; faculty privileges	Technology Transfer laws; diversity guidelines
Dominant Ideology	Advancement of Knowledge	Creating the Future

Source: Brint, 2005: 48.

Challenges for Irish HEI Leaders & Administrators

Irish HEIs face the key challenge of balancing the tension between maintaining high teaching standards and developing sustainable collaborative efforts, while remaining true to their scholastic traditions. HEI leaders will have to develop a clearer focus on their mission and objectives and will have to develop specific policies and procedures to facilitate the implementation of their third-stream activities. The

Lambert Review in the UK (2003) highlighted the need for effective leadership and the development of commercialisation skills and awareness among HEI leaders and administrators. Further, the review highlighted the importance of regional co-operation and the potential of shared service models for technology transfer. HEIs will need to consider which third-stream activities their institutions will pursue, given their current areas of international expertise and knowledge competencies. This also will entail examining the best means of co-operation with other HEIs and industry regionally, nationally and internationally. In the light of these challenges, Irish HEI leaders and administrators will have to think, act and manage strategically. Short-term initiatives or maintaining the *status quo* are no longer an option, if HEIs are to play their part in the knowledge-based economy.

While it is often the case that the potential exists for commercialisation, it is frequently undermined by a lack of mutual understanding (Leydesdorff, 2000). Commercialisation activities should be endorsed and promoted as a valuable outlet for research results (Siegel *et al.*, 2003; Lambert, 2003). Propensity to commercialise, as well the development of knowledge and realistic expectations as to the nature of the technology transfer process, are best developed by Technology Transfer Officers working on the ground with academics.

A critical challenge facing HEI leaders is that of persuading faculty members of the strategic importance of commercialisation. King's (2005) study[8] of 1,000 research academics, for example, demonstrates that Irish HEIs have a lot of ground to make up. Half of the respondent researchers in his sample have never been involved in collaborative research with industry, while 70% of the respondents had never been involved in any commercialisation activity resulting from academic research. The validity and role of commercialisation activities in terms of delivering research findings for the public good therefore should be communicated and highlighted as an inherent part of the university's mission (Den Hertog *et al.*, 2003). For HEI leaders, this means attaching value to these activities and, in practice, means putting in place support and incentive systems while balancing these activities with prudent governance structures. Moreover, it means HEIs actively educating both business and society in general in terms of their core missions, one of them being an active and valuable contributor to Ireland's knowledge-based economy.

Structure of the Book

The rationale for this book is rooted firmly in the context of the developments and challenges highlighted throughout this chapter. **Chapter 2** reviews how various technology transfer structures have

---

[8]   As reported in Downes, J. (2005).

evolved overtime, prior to examining the value of specific structures –
M-form, holding company, and the matrix structure. **Chapter 3**
examines the roles and activities that are critical to successful
technology transfer interventions. In **Chapter 4**, we explore some of
the theory and research related to technology transfer, before
considering technology transfer mechanisms, including licensing,
patenting, and spin-offs. **Chapter 5** considers the barriers and
stimulants to technology transfer, and highlights the difficulties in
measuring and evaluating technology transfer outcomes. In order to
illustrate some of the issues being discussed, as well as to highlight the
diversity of institutional approaches to technology transfer, **Chapter 6**
provides evidence from three best practice case studies: the University
of California in the US, the Hebrew University of Jerusalem in Israel
and the case of regional co-operation in Finland. In the final chapters,
Chapters 7 and **8**, we attempt to bring together the 'varieties of
excellence' we have reviewed to form a model to guide the strategic
management of technology transfer. This model focuses on the
structure, activities, staffing, policies, mechanisms and evaluation
procedures we believe to be critical to successful technology transfer
initiatives. **Chapter 8** also draws attention to the imperative of
researcher motivation and a general ethos of commercialisation in
sustaining this model. The book concludes by reviewing the lessons
that may be learned and the gaps or opportunities that may be
identified from applying the model to Irish HEIs.

2

ORGANISATIONAL STRUCTURES FOR TECHNOLOGY TRANSFER

The question is not whether increased university-industry
collaboration can yield desirable outcomes for all concerned:
clearly, it can and often does, the question is how?
J. David Roessner, Georgia Institute of Technology[9]

INTRODUCTION

In this chapter, we examine the different organisational structures that may facilitate the effective management of technology transfer in a HEI context. We begin by exploring the evolution of technology transfer structures and outlining common organisational arrangements for technology transfer. Then, using three in-depth case studies of Technology Transfer Offices (John Hopkins University, Duke University, and Pennsylvania State University), the relative merits of three organisational forms for structuring technology transfer (matrix, H-form, and M-form) are assessed. We conclude by summarising the key issues drawn from the discussion.

TECHNOLOGY TRANSFER OFFICES

While the structural problems faced by HEIs in developing and sustaining linkages with business and industry are similar, responses vary by country. One constant, however, is that greater attention is being paid to governance and management issues within universities. Many universities have attempted to formalise technology transfer by establishing Technology Transfer Offices (TTOs). These offices play a crucial role in developing the networks for exchange between university researchers and potential users of their research results in the private sector (Graff *et al.*, 2001: 27). In Irish HEIs, the title Industrial Liaison Offices (ILOs) has been used to describe the unit

---

[9] Roessner (1996).

responsible for technology transfer. Internationally, different titles are used, such as Technology Transfer Offices and Technology Licensing Offices (OECD, 2002; 2003), as well as Corporate Liaison Offices, Office of Corporate Relations or Technology Transferring Points (Graff *et al.*, 2001; Sanchez & Tejedor, 1995). Despite these different labels, the activities of these offices are essentially consistent, in that they focus on key elements defined by the OECD (2003: 37) as: 'facilitating the identification, exploitation, and defending of intellectual property' or more simply as captured by Scott *et al.* (2001) 'to manage the transfer of intellectual property to third parties'.

Increasingly, HEIs face a widely perceived contrast between excellence in science and technology and a relative weakness in exploiting them to economic advantage (Grady & Pratt, 2000). This has lead to a quest to the find the most efficient means to organise technology transfer activities. Yet resource constraints, and the dynamics of growth and change, make this a difficult task. Commercialisation activities in the HEI sector have increased substantially in the last five years. In the UK, for example, many HEIs only created TTOs in the late 1990s; now 80% have at least one dedicated person, while average staff numbers are still rising by almost 25% per annum (see **Figure 2.1**).

Figure 2.1: UK Technology Transfer Offices

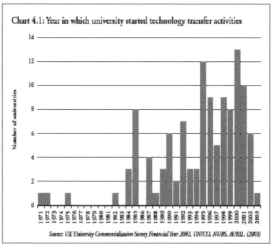

What is evident in an Irish context is that those HEIs that undertake very high levels of R&D and have more comprehensive technology transfer activities have moved to create dedicated Industry Liaison or Technology Transfer offices (ILOs or TTOs), while in many other institutions, such as many of the Institutes of Technology, technology transfer is given a low priority (Forfás, 2005). The role of such TTOs includes liaising with scientists, researchers and industry, and with licensing and patenting attorneys. In most OECD countries, TTOs are small operations, with fewer than five full-time employees and their work programmes often extend beyond their strict role in technology transfer (Forfás, 2004a: 34). The formation of a centralised TTO usually introduces legal formalisation and creates an institutional focal point for the flow of technologies out of the university and into industry (Graff *et al.*, 2002).

Variety of Models & Organisational Arrangements

Evidently, a range of technology transfer procedures and policies, as well as organising mechanisms for technology transfer offices, have been developed in response to legislation and market opportunities (Bercovitz *et al.*, 2001). Structural options, reporting levels and levels of autonomy open to TTOs are normally heavily shaped by institutional norms and path dependencies. The right mechanisms may depend upon the scale and substantive focus of the collaboration, as well as the mission, objectives and strategic direction set at an institutional level.

TTO Models

The requirement for increased attention to commercialisation activities is generally manifested through some form of organisational development and the setting-up of specialised internal support structures for the development of relations with industry (Hernes & Martin, 2000). The various institutional responses include:

♦ Setting up a single centralised TTO (smaller universities).

♦ Having a number of TTOs associated with different schools or departments (multi-campus universities, such as the University of California).

♦ Outsourcing technology transfer to an external agency, which may have several client organisations (OECD, 2003: 80).

Previous research has identified a more detailed typology of the institutional arrangements in place for exploiting intellectual property. The several models employed by universities are listed below:

♦ **Model 1:** One entry point for industry; the company is then directed to the appropriate office (University of Georgia Technology Office).

♦ **Model 2:** One-stop-shopping for faculty and industry pertaining to technology transfer, industry-sponsored research, material transfer agreements, industry pre-clinical and clinical trials (John Hopkins University).

♦ **Model 3:** Separate offices for technology transfer and industry collaborations, clinical trials, etc (found in many universities in the US).

♦ **Model 4:** Combined offices for technology transfer and industry collaboration in one office (Stanford University).

♦ **Model 5:** One office responsible for all university-industry interactions, including purchasing (Duke University).

♦ **Model 6:** Traditional TTO, managing inventions, patent/copyright filings, marketing, licensing, start-up companies, material transfer agreements, and building relationships with industry with an emphasise on in-state companies (found in many universities).

While many of these models can be made to work at any HEI, they are not necessarily mutually exclusive. Normally the first model to develop at most institutions is an office that provides only traditional technology transfer services (Model 6). For campuses with a small technology transfer office (less than 10 inventions per year), it is best to start with modest expectations and to provide limited services. In this instance, the sponsored research office would handle industry-sponsored agreements.

Organisational Arrangements

The institutional arrangement chosen at each HEI tends to reflect its history and research base. Hence, arrangements will differ in their degree of centralisation, in accordance with the general culture prevailing at the institution and its size (Martin, 2000). A decentralised approach is natural in the case of a multi-campus university distributed over a number of locations – for example, the University of Sau Paulo – while a centralised approach is evident at the Hebrew University of Jerusalem, where an external body, YISSUM, was created specifically to deal with technology transfer. In the latter case, rights and obligations have been laid down in the university management regulations, which YISSUM must strictly adhere to, as outlined in **Table 2.1**.

Table 2.1: Interface Structures: YISSUM at the Hebrew University of Jerusalem

YISSUM is a non-profit, wholly-owned subsidiary of the Hebrew University of Jerusalem (HUJ). Its aim is to manage all research-related relations with industry – for example, with respect to contract negotiation, patenting, licensing, venture capital and the development and management of spin-off companies. This independent legal status allows YISSUM both to protect the university from undue external interference and to obtain managerial flexibility in the daily running of operations and projects.

HUJ directly controls YISSUM. HUJ has a consistent body of rules and regulations – for example, determining the amount of time a researcher can afford to spend on external activities, incorporation of an obligation to disclose outside economic interests and research of potentially-commercial interest. IP rights of R&D results remain within the university, and YISSUM is the entity in charge of the commercialisation of R&D and technical services. There is a strict policy with regard to researcher involvement in business – their equity stake may not exceed 10%.

THE EVOLUTION OF TTO STRUCTURES

Traditionally, universities appointed Technology Transfer Officers, usually with strong industrial backgrounds, to maintain links with industry. R&D was seen as a linear process. TTOs tended to be sited on the periphery of the university campus and were usually left to their own devices – preparing information brochures, advising individual academics and companies, and visiting companies to keep abreast of industrial developments. However, the developments highlighted in **Chapter 1** suggest that TTOs are increasingly required to be more proactive, more interested in a professional technology role and more concerned with identifying and exploiting intellectual property. The extent and complexity of university and business relationships has meant that TTOs are taking on an increasing number of tasks as universities' engagement with their wider community has developed. This, in turn, has necessitated the development of appropriate structures to facilitate and accommodate growing activities and demands for services (Shattock, 2001).

There is no single universal organisational model for third-stream activities. Some institutions include knowledge transfer and technology transfer activities as part of the functions of their TTO, while others keep the two activities separate and have established specialised companies to manage technology transfer. The appropriate structure will vary depending on the needs of local business, the mission of the university, and the focus of the local economy.

Technology Transfer Offices

Anecdotal evidence from successful TTOs suggests that they develop from a single point of contact towards the creation of a number of independent specialised units, focusing on specific activities with external actors. Trends suggest that TTOs have expanded from one-person operations to offices with several specialists in various aspects of technology transfer working under a director (Jones-Evans, 1999; Scott *et al.*, 2001)

Case Management Approach

Another approach evident from the US is what has been termed 'case management' or the 'cradle to grave' approach. This structure involves specialists in various research areas conducting the full range of technology transfer activities. In this system, typically one person is responsible for the actions required for a particular case, from disclosure through patenting, and sometimes beyond. This approach offers the advantage of centralising awareness and coordination with all major aspects related to a particular intellectual property.

The case management style was common in many of the TTOs studied by Allan (2001) in his review of best practice.

Funding of TTOs

The issue of the appropriate structure and management of technology transfer is linked inherently to the level of funding that TTOs receive from their HEI and also from external agencies and sources. This is a key issue in terms of developing an appropriate structure, given the institutional context within which TTOs reside. With the growth of industrially-related research, the issue of the appropriate level of overhead charges, or indirect costs that need to be levied, became more important, especially as state funding levels declined (Shattock, 2001). For example, at Warwick, the TTO was merged with the part of the Finance Office that was responsible for the financial side of research grants and contract administration, with a view to creating a Research & Development Services Office. This office is responsible both for negotiating overheads and accounting for research contract income. The office is also responsible for the overall central research management function, which had grown enormously because of the extent to which success in research (often equated with the size of overall external research income) was defining the long-term status and profile of the institution. This structure only represents a small element of what is required today.

External Support for Technology Transfer Offices[10]

In many countries, governments have made pro-active efforts to support the initiation and development of TTO at various institutions. In 2002, for example, the German Ministry of Research & Technology launched a multi-million Euro programme to assist HEIs in hiring external services for licensing and prosecution of IP (Gering & Schmoch, 2003). In France, the Innovation Law of 1999 provides for the strengthening of TTO structures at HEIs, notably through the creation of departments for commercial and industrial service activities.

Some countries are experimenting with regional or sector-based offices – for example, Denmark, UK, and Germany. This has been driven by the logic that smaller universities may benefit more, either by aligning themselves with the TTOs of larger universities, or by forming collective TTOs among themselves in order to capture economies of scale.

The idea of some level of shared costs and regional alliances seems sensible. In Ireland, several inter-institutional alliances are already in place, such as the Atlantic University Alliance, the Technology Transfer Initiative and TecNet (Forfás, 2004a). These alliances have evolved around collaboration in research and educational programmes. To date, however, no inter-institutional function has been created to deal specifically with elements of technology transfer or research commercialisation activities. In the UK, some Public Research Organisations, with government support, have developed a partnership to pool resources and to increase the rate at which they market their intellectual property in the health and life science fields. An example of one such regional partnership is provided below in **Table 2.2.**

---

[10]   This issue is explored in greater depth as part of the key issues relevant to the development of a strategic approach to technology transfer (**Chapter 7**). **Chapters 2 to 5** provide a rationale for such structures and place their development in the context of activities conducted by TTOs, mechanisms for technology transfer and barriers to commercialisation.

Table 2.2: Regional Partnerships for Managing Intellectual Property

A network of hub organisations is being established across England to manage IP coming out of hospitals and health organisations. These hub organisations receive public funding from the Department of Trade & Industry, the Department of Health, and Regional Development Agencies. Concentrating expertise in a hub, which can be centred in one or more locations in each region, allows for the efficient management of IP through the concentration of resources and recognises that most NHS organisations are too small to justify such a dedicated activity. Hubs will normally be found close to university medical schools and will work closely with university TTOs. This is important in terms of sharing knowledge and expertise when dealing with IP resulting from joint research between hospitals and universities.

The prototype of this kind of partnership is centred around Manchester, where three NHS Trusts worked in partnership with four universities (Manchester, UMIST, Salford and Manchester Metropolitan). Their MANIP partnership (Manchester Intellectual Property) received funding from the DTI's Biotechnology Exploitation Platform (BEP). IP is identified, evaluated and an exploitation route agreed. Much of the IP arises from joint work between the NHS Trusts and the universities, and the route to exploitation is managed by the partner universities' TTOs or by one of the partner Trusts.

Source: Adapted from DTI (2004); OECD, 2003.

Optimal Institutional Arrangements for Organising TTOs

The structure of TTOs is heavily contingent on a number of factors:

♦ A legal environment dictating that HEIs can claim title to inventions tends to provide a greater incentive to develop and invest in technology transfer structures.

♦ The HEI's degree of institutional autonomy and the existence of laws and regulations that require the TTOs to adopt alternatives to in-house operations. For example, in the US, many public and state chartered universities have established arms-length institutions (foundations), because they generally benefit from the immunity from prosecution granted to state governments. In Japan, national universities are not autonomous and TTOs have been established as separate private entities. In Israel, the TTOs at the Weizmann Institute (Yeda) and at the Hebrew University (YISSUM) were established as fully-owned subsidiary companies to allow PROs to earn revenue and hold equity in spin-off companies.

♦ Clear institutional rules regarding equity participation in spin-off activities. For example, rules in many European countries prohibit public HEIs from having equity participation in spin-offs. Recently, the UK changed a law prohibiting HEIs from keeping revenue from commercialisation. Previously, licensing revenue was transferred to the Treasury. Korea amended its legislation in 2001 to allow TTOs in public universities to become legal entities, thus making it possible for them to appropriate financial returns from licensing.

In general, the appropriateness of one institutional arrangement or another depends on:

♦ The context in which the PRO operates.

♦ Its status as a private or public institution.

♦ The amount of government funding it receives.

♦ The size of its research portfolio and fields of specialisation.

♦ Its geographical proximity to firms and inclusion in innovation networks.

♦ Its funding capacity (OECD, 2003).

Increased attention has been directed at technology transfer and the requirement for TTOs to have a clear structure and reporting relationships (Shattock, 2001). While there is no optimal structure of TTOs, a number of guidelines taken from best practice can facilitate the decision process (see **Table 2.3**).

Table 2.3: Guidelines for Developing an Appropriate Organisational Structure for TTOs

♦ Ensure clear lines of reporting and responsibility.

♦ Ensure close interaction between commercialisation and research activity.

♦ Typically, the TTO director reports to a Vice President, Vice Provost or Vice Chancellor for research (Siegel *et al.*, 2003).

♦ Improve communication and teamwork among HEI personnel, consider co-location of key offices.

♦ Co-ordinate the efforts of various offices to support researchers.

♦ The TTO must have a prominent position in the university structure and among top management.

♦ Executive Committee to oversee commercialisation activities.

♦ Development of specialists (by TTO activity or by research area).

♦ Although the functions of research and technology transfer are related, the two offices have very different goals and should be administratively independent.

Sources: OECD, 2003; Shattock, 2001.

VARIETIES OF ORGANISATIONAL STRUCTURE

Implementation structures and reporting structures have been the subject of much debate within HEIs and, as a result of historical reporting lines, can be highly politicised. A component of effective management is a simple management structure. It is not uncommon for HEI presidents/directors to have dozens of direct managerial reports, both academic and administrative. Reporting structures matter less in collegial, committee-based systems, but the more executively an HEI is run, the more reporting lines need to be rationalised and clarified. This has led many HEIs towards sweeping changes in the number of schools, faculties or departments, in order to cut back the number of reporting lines in the HEI administration – for example, recent changes at Trinity College, Dublin (see also Lambert, 2003).

Rather than simply looking at TTOs' activities, interest has shifted to understanding the reasons for inter-institutional variations in the range and efficiency of technology transfer activities. There is a requirement to consider how organisational structure mediates the relationships between the inputs that give rise to intellectual property and the level and forms by which the HEIs generates revenues. Technology transfer activities, therefore, can be seen to be shaped by the resources, reporting relationship, autonomy and/or the incentives provided to TTOs. The monolithic set of structures that HEIs previously employed to relate to industry are now seen as no longer relevant and, increasingly, successful HEIs are adopting a range of organisational instruments to relate to the industrial world. Extensive research in this area concluded that much of this development is path-dependent and related to the specific circumstances of each university (Bercovitz *et al.*, 2001).

Organisational Structure as a Determinant of Academic Patent & Licensing Behaviour: Evidence from the US

Technology transfer outcomes may depend on organisational practices that potentially attenuate palpable differences in the motives, incentives, and organisational cultures of the players involved in this process (Siegel *et al.*, 2003). Building on this rationale, Bercovitz *et al.*, (2001) examined the organisational structure of TTOs as an independent variable that accounts for inter-institutional variances in technology transfer and patenting. Although focusing on merely three universities – the John Hopkins University, Duke University, and Pennsylvania State University – these universities are noted as being major research universities (Thursby & Kemp, 1999). Furthermore, such depth of analysis provides invaluable and detailed insights into the merits of alternative organisational structures and thus may serve to inform such considerations in alternative contexts.[11]

A Review of Organisational Structure Types

Bercovitz *et al.* (2001) drew on a well-established literature to examine organisational structures (see **Table 2.4** to appreciate the content of the literature and to put the preceding discussions in context). These organisational structures offer insights for analysing universities. Considerable diversity exists in technology transfer procedures and policies, as well as in the organisation of technology transfer procedures and policies and also the organisation of technology transfer and intellectual property offices. This diversity maybe viewed as a natural experiment, in which the various actors search for the most efficient means to organise their activities in a relatively uncharted domain.

Methodology & the Universities for Bercovitz's Study

The sample of universities used for this study included Johns Hopkins University, Duke University, and Pennsylvania State University. Data was collated from a series of round-robin interviews with technology transfer personnel, faculty and research administrators at each university. In total, 21 interviews were conducted, and secondary data in the form of policies, organisation charts and databases was also used. The researchers also examined the historical context of the universities, as this was deemed to: 'set the stage for the organisational structure of the technology transfer operations' (Bercovitz *et al.*, 2001). (See **Table 2.5** for a comparative summary of the three universities.)

In the past 20 years, distinct, well-staffed technology transfer organisations have emerged at each institution and, according to efficiency scores developed by Thursby & Kemp (1999), each of these universities operates an efficient technology transfer office. That is, each of these universities operates on the technology-transfer production frontier and can be considered 'best practice institutions'. In all the universities, it was found that institutional history, culture and norms of behaviour, while not the sole determinants of the structure of the university technology transfer efforts, played an important role in the universities' approach to technology transfer.

---

11   The strategy-structure linkage employed is a partial approach to examining the full set of factors that account for inter-institutional variations. Bercovitz *et al.* (2001) acknowledge this, noting that other determinants may include a) the level and composition of sources of a university's research funding (federal/state/ industry/foundation/other); b) quality of research performed by the university; c) the university's commitment to patent and licensing.

Table 2.4: Competencies & Characteristics of Alternative Organisational Structures

Organisational Structure	Characteristics	Locus of Decisions	Information Processing	Co-ordination	Incentive Alignment (across units)	University
Functional or Unitary (U-Form)	Centralised, functionally departmentalised structure	Small group at the top of the hierarchy; University administration	Limited by HQ size; the need to funnel decisions through top management creates a bottleneck	Relatively strong among sequential work units given vertical control	Difficult to create unit-level incentives compatible across units and in-line with organisational goals	
Multi-divisional (M-form)	Semiautonomous operating divisions along customer/ product/ geographical lines	Segmented decision-making responsibility	Decentralised decision-making leads to a higher overall information processing capability within units	Strong central body allows for moderate top-down co-ordination across units	Strong unit level incentives; sub-goal pursuit is problematic but tempered by stronger organisational ties	**Pennsylvania State University**
Holding Company (H-form)	Divisional approach, weak central office	Division	Decentralised decision-making leads to a higher overall information processing capability	Weak central body allows for moderate top-down co-ordination across units	Strong unit level incentives; sub-goal pursuit is often problematic but tempered by stronger organisational ties	**John Hopkins University**

Organisational Structure	Characteristics	Locus of Decisions	Information Processing	Co-ordination	Incentive Alignment (across units)	University
Matrix Structure (MX-Form)	Combines two or three more divisions of function, product, customer or place	Structure in which individual or sub-units are responsible for multi-dimensional functions	Multi-dimensional responsibilities may tax information processing capacity within units	Dual dimension responsibilities drive co-ordinated action	Dual incentives: functional and product incentives are integrated to reflect organisational goals	**Duke University**

Table 2.5: A Comparative Summary of Three Universities

	JHU	Duke	PSU
Type of institution	Private	Private	Public
Founded	1876	1924	1854
US Patents 1980-90	153	56	6
US Patents 1991-97	201	134	96
Rank in academic R&D expenditures	8th	26th	10th
1998 Invention disclosures	228	113	190
1998 Patents (AUTM)	76	37	25
1998 Active licenses (AUTM)	149	45	68
1998 Licensing income	$5,513,284	$1,318,680	$2,012,584
Spin-offs	5	1	5
Efficiency rating (Thursby & Kemp)	100%	100%	100%
Organisational structure	H-Form	MX-Form[12]	M-Form

Source: Bercovitz et al. *(2001).*

H-form Technology Transfer Structures at John Hopkins University

John Hopkins University is geographically-spread throughout Baltimore, reflecting a historical and spatial differentiation, with a large degree of autonomy among its constituent units. The Homewood main campus is eight miles downtown from the School of Medicine. The Peabody Music School is located in midtown, while the Applied Physics Lab is located in the suburbs. Each of these units has a technology transfer office. These offices are loosely connected by an overarching governing body, consisting of the University President and Board of Trustees.

Independent sponsored research units, which maintain a co-ordinated reported system, also exist at each site. Each relevant division supports a TTO. No systematic allocation of licensing revenues exists to provide an incentive to the offices. Thus, John Hopkins University's technology transfer operations appear to be organised in an H-form structure.

The Homewood Office of Technology Transfer, created in 1976, overseas the Schools of Arts & Sciences, Engineering and Public Health. The OTT operates under the auspices of Vice-Provost for Research and is responsible for technology transfer activities. Peabody

---

12 Operates a collapsed matrix structure, which is a version of the matrix structure in which unit managers hold primary responsibility along both organisational dimensions.

has aggressively combined its traditional strengths in music with an entrepreneurial vision to develop products related to multimedia composition and performance and digital audio processing systems.

Figure 2.2: John Hopkins University's Organisational Structure

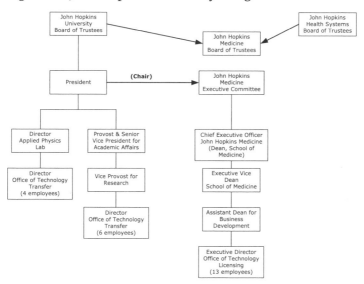

The Office of Technology Licensing (OTL) at the School of Medicine is located 10 minutes from the Medical School Campus. Created in 1986, it operates under the School of Medicine's research administration department. OTL works closely with the Office of Corporate Liaison, which is responsible for cultivating relationships, and the Research Administration Office, which is responsible for negotiating research agreements. The Applied Physics Laboratory was added to the Office of Technology Transfer (APL-OTT) in September 1999. The new office was created with the following goal: 'in addition to a typical technology push approach, the office will also establish a market-pull approach: identifying the needs of the commercial sector that match unique capabilities and multi-faceted strengths of the Laboratory' (APL press release, 1999).

In summary, John Hopkins University's technology transfer effort is consistent with an H-form structure, with four decentralised technology transfer operations. Each unit is funded largely by, and reports to, divisional offices and there is limited central administrative control. While there are plans to provide incentives for the offices at OTT and OTL to share licensing revenues, there is no formula in place.

M-form Technology Transfer Structures at Pennsylvania State University

As a land grant institution, Pennsylvania State University (PSU) began with a mandate of service to the public. It has continued a commitment to the development of practical sciences for the purpose of enhancing economic growth and general well-being. PSU has evolved as an M-form organisation. In 1987, the board of trustees endorsed a set of initiatives that included strengthening the university's research and technology transfer facilities; developing programmes to assist faculty, students and staff in entrepreneurial activities; and building recognition and support for an active role in state economic development. These initiatives catalysed a reorganisation of the University's Office of the Vice-President of Research, the creation of the position of Associate Vice President for Research and the establishment of an Office of Technology Licensing. Reflecting the University's commitment to technology transfer, the new associate vice-president position also included responsibility for administration of sponsored research activities, thus linking administratively sponsored research with licensing activities.

Figure 2.3: Pennsylvania State University's Organisational Structure

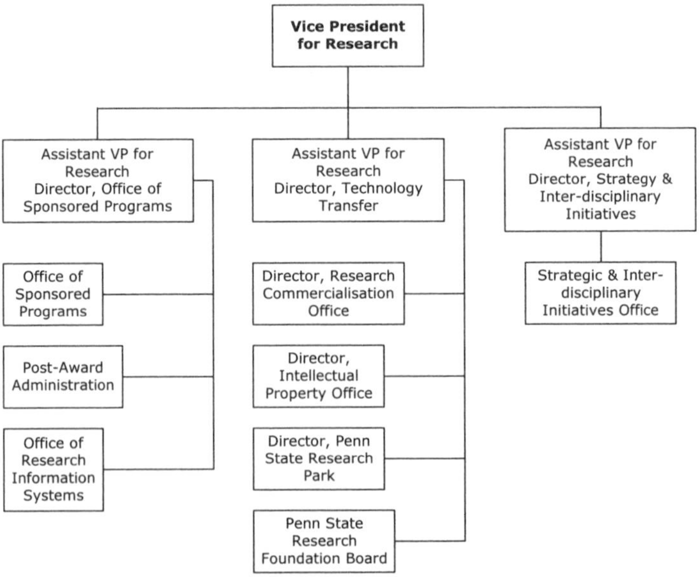

In the mid-1990s, the consensus at PSU was that technology transfer, although increasing, had not generated the financial outcomes or state economic development contributions commensurate with the university research base. In response, technology transfer was again reorganised. A task force recommended splitting responsibility for sponsored research and licensing/technology transfer. This led to the creation of assistant vice-presidents for both technology transfer and research and strategic initiatives, each reporting to the Vice-President for Research. In addition, a new position of Director of Technology Transfer was established in 1995 to support technology transfer at the Hershey Medical Centre, located 90 miles away. Over the past five years, Penn State's organisation of its technology transfer activities has moved closer to the M-Form.

Matrix Technology Transfer Structures at Duke University

Duke University, in contrast to John Hopkins and Penn State, is more integrated both in form and geography. The Medical School is located on the university quadrangle and is an integral part of the university. From the beginning, there has been one human resources, payroll, accounting and administration services for all of the university. Duke's Office of Science & Technology (OST) has evolved as a collapsed matrix structure. The head of OST, the Vice Chancellor of Science & Technology Development, reports directly to the Chancellor of Health Affairs, who in turn reports directly to the university president. Within OST, there are four Associate Directors. Three of these individuals have PhDs in the sciences, while the fourth holds an MBA. In the collapsed matrix, each associate director has primary responsibility for one or two facets of the university-industry technology transfer initiative, broken-out by sponsored research agreements, patents and license administration, spin-off activities, materials transfer agreements and corporate partnering and business development. Simultaneously, each of these individuals oversees a specific technology area – medicine, medical devices, biotechnology (bio-medical and agricultural) or chemical, physical sciences and engineering. Each of the Associate Directors manages patenting and licensing within his or her technology area.

The OST is funded through remitted shares of intellectual property revenues generated, receiving 0.5% of the university indirect charges for sponsored research. According to the university's patent policy, the OST receives 10% of licensing royalties, after expenses. In practice, the share captured by the OST is closer to 30% of licensing revenues, as the University has chosen to direct to the OST the 20% share of royalties designated to 'provide research support in the University as determined by the President upon the advice and counsel of the Chancellor for Health Affairs' (Duke University Patent Policy). The Office receives revenues based upon both licensing royalties and research funding. Given the higher risk associated with the uncertain yield and timing of licensing royalties, the Associate Directors have sufficient incentives to pursue either mechanism.

Figure 2.4: Duke University's Organisational Structure

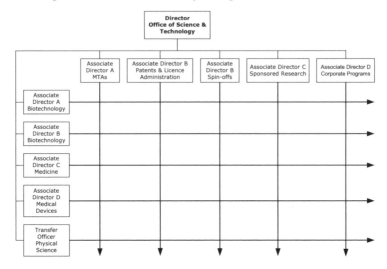

The OST staff are housed in several different buildings across, as well as off, campus. Co-ordination is maintained through frequent phone calls, e-mails, meetings and use of the common technology transfer database in which all invention disclosures, patenting efforts and licensing activities and research agreements are tracked.

TTO Performance & Organisational Structures

Interestingly, Bercovitz *et al.* (2001) provided a set of testable propositions linking organisational form with performance ramifications for technology transfer. The hypothesis centred on each of the performance attributes introduced in **Table 2.4**: information processing capacity, co-ordination capabilities and incentive alignment properties. The key findings are summarised in **Table 2.6**.

Table 2.6: TTO Performance & Organisational Structures

Performance Attribute	Description	Findings
Co-ordination capacity	Capacity to orchestrate the activities of different work units. Relevant units include any entity that has interaction with business – involved with sponsored research, IP management, student hiring	More likely to leverage relationships and share information through a matrix structure (Duke), than M-form (Penn State), with the H-form (JHU) least likely.
Information-processing capacity	Throughput or yield achieved by the office (average number of technology transfer transactions) – invention disclosures, licenses, sponsored research agreements (normalised by staff size)	Yield is greater through M-form (Penn State) and H-form (JHU) and slightly less through the matrix structure (Duke)
Incentive alignment	Extent to which TTOs are awarded for various mechanisms – leveraging sponsored research revenue against licensing revenue	Within the bounds of evolving legal precedent, the tendency to trade-off royalty revenues for sponsored research support is greatest at Duke (matrix), followed by Penn (M-form) and least likely at JHU (H-form); little cross unit alignment with H- or M-form

Adopted from Bercovitz et al. *(2001).*

The results presented are based on detailed case studies of very dissimilar universities in terms of their organisational form, strategies for establishing intellectual property rights and in securing revenues from these rights. Bercovitz *et al.* (2001) conclude by acknowledging that universities have different histories, capabilities and resources and thus evolve structurally in different ways, but that each structure in turn will affect performance in a visible manner:

♦ **John Hopkins University**, for example, aligns with the H-form and, reflecting the historical autonomy of the individual units, is organised as multiple decentralised technology transfer units with limited central control. Within the structure, there are unit-level incentives, unit-level information processing capacity, although strong unit-level yields.

♦ **Penn State University** has created an M-form organisation, with a centralised administration office and decentralised units. This organisational form, while similar in information processing capacity, offers slightly more across unit co-ordination and incentive alignment competencies than the H-form.

♦ **Duke University** has chosen to organise its activities within one office *via* a collapsed matrix form, reflecting its more unified history. Although this organisational form sacrifices local yields, it results in greater across unit co-ordination and incentive alignment. Consequently, Duke provides a better interface mechanism for interaction with business.

THE LESSONS FOR HEIs

Based on this chapter, and the overview of three differing organisational structures, some key lessons can be drawn for HEIs:

♦ Technology transfer structures at HEIs are **complex** and **path-dependent**. Each organisational structure can only be understood in the context from which it emerged. Historical appreciation of this is crucial, particularly in estimating potential cultural resistance should structural alterations be attempted.

♦ Organisational factors and **structure** can **affect** technology transfer **outcomes**.

♦ In the context of HEI technology transfer, the **degree of co-ordination** among licensing, sponsored research, and related technology transfer units can determine the extent to which these units integrate their activities and share information related to business needs and opportunities.

♦ In terms of generic organisation structure, the **most effective** seems to be the **matrix** structure. This enables the highest incentive alignment, as returns are generally a function of activities undertaken along both axes of the matrix. The matrix structure, although not the most efficient in terms of narrowly-measured throughput or yield, is deemed to be the most effective structure in terms of its overall capacity to co-ordinate and incentivise the university-wide interface with business.

♦ **Leveraging ability**, the trade off between royalty rate/licensing fees and sponsored research dollars will also be greatest when technology transfer activities are structured in a matrix organisational form.

♦ HEIs have to consider the appropriate **title** to be given to the office that has responsibility for technology transfer and which is reflective of key stakeholder interests.

3
TECHNOLOGY TRANSFER ROLES, ACTIVITIES & RESPONSIBILITIES

INTRODUCTION

In this chapter, we focus on the roles, activities and responsibilities of those involved in technology transfer. The chapter begins by discussing the historical context of TTOs and their main roles: switchboard services, network development and technology transfer, and managing IP activities. The chapter then examines the multiple responsibilities of TTOs in terms of their key stakeholders and in maintaining true to the HEIs' mission. The chapter concludes by examining staffing levels of TTOs, as well as the functions and skills said to be crucial to successful technology transfer.

HISTORICAL CONTEXT

TTOs traditionally play several important roles in the process of technology transfer, including information broker, science marketer and catalyst for academic entrepreneurs (Fassin, 2001). Several leading international research universities, including the University of California, Stanford, MIT, and the University of Wisconsin, established their respective TTOs over 30 years ago. The Research Corporation, founded in 1912 by a University of California-Berkeley faculty member to sell rights to use patents of several affiliated universities, was a predecessor to these contemporary offices (Mowery *et al.*, 2001).

While the legislative context varies across the world, the US model of commercialisation, and its association with economic growth, has been the one most countries have attempted to emulate. The positive effect of the Bayh-Dole Act in encouraging and providing an incentive for commercialisation is well-documented. A survey of 106 university and government research laboratories and TTOs conducted by Castillo *et al.* (2000) found that 78% of the TTOs surveyed were created after the passage of the Act. The Association of University Technology Managers (AUTM) reports that, in 1980, only 25 offices existed, while a decade later there were some 200 offices (reported in Mowery *et al.*, 2001). Similarly, an historical analysis by Stevenson (2003) has documented further developments in the commercialisation activities and measurements used by HEIs. Despite their contemporary significance, it has been noted that very little research has been conducted into the role and value of TTOs (Scott *et al.*, 2001).

ROLES & ACTIVITIES OF TTOs

Definition of the various roles that TTOs should pursue diverge between those who favour a narrow role for TTOs – primarily as a switchboard – and those who favour a broader role – of helping two-way communications between HEIs and the outside world – for example, identifying curriculum development needs. Yet this is a simplistic dichotomy; in reality, the function of TTOs varies from HEI to HEI and from country to country.

Forfás (2004a) notes that the key types of technology transfer supports provided by Irish HEIs tend to be:

♦ The protection and licensing of IP.

♦ Provision of incubator centres on campus.

♦ Setting up spin-out firms supported by the research institution.

♦ Research for industry, either directly on contract or indirectly through collaborative research programmes co-funded by the State.

♦ Provision of advice, information, consultancy and training programmes.

♦ The supply of graduates, post-graduates and doctorate-level researchers.

Overall, the main function of a TTO is to provide a formal, above-board, and relatively effective mechanism for those researchers who wish to commercialise their ideas (Graff *et al.*, 2001: 14). In this sense, the role of TTOs can be broken down into four main areas:

♦ Switchboard services.

♦ Network development.

♦ Technology transfer.

♦ Managing IP activities.

Switchboard Services

EIMS (1995) refers to technology transfer as a formal function of the HEI, involving the management of the interface between academia and various external institutions. HEIs are complicated institutions, and businesses can find it very difficult to find their way around them. SMEs, in particular, can be put off if there is no obvious point of entry to the HEI's resources. This is the traditional role of the TTO, serving as a signpost offering switchboard services and directing industrialists who are seeking the most appropriate expertise within the HEI (Jones-Evans, 1999). A key role of the TTO involves promoting HEI-industry interaction. The goal of the office is to lower the entrance barrier for the external business world and to complement existing informal direct contacts between faculty and industrial representatives.

Network Development

Increasingly, the pro-active role that HEIs can play in developing strong linkages with industry, particularly through mechanisms such TTOs, has been highlighted (Scott *et al.*, 2001). TTOs facilitate this through providing opportunities for enterprising staff to develop ideas and contribute to the economy *via* knowledge transfer and spin-off companies. Best practice examples of TTO activities are largely US-based, where TTOs have extensive resources and staff to facilitate commercialisation and HEI support of these activities is taken as a given (MIT, University of California). The reality in most other countries is that, up until recently, there is 'little information to suggest that ILOs are undertaking a pro-active role ..., at most, industrial liaison offices at higher educational institutions are merely providing marketing services for their parent organisation' (Jones-Evans *et al.*, 1999: 49).

Jones-Evans *et al.* (1999) conducted a comparative study of TTO roles and activities in Ireland and Sweden, using interviews with key individuals involved in the process of technology transfer in both countries. Their results indicated that the role of TTOs should be to provide a strategic focus for HEI-business collaboration, although how this became manifest differed by country. In Sweden, the TTO provides a gateway to a network of technology transfer organisations. In Ireland, the development of both internal and external formal networks by TTOs was judged to be in a primitive state.

Management of Technology Transfer

In general, the function of the TTO is to manage the activities that occur as part of the collaborative efforts between HEIs and business. The TTO manages the process of invention disclosure and evaluation, takes decisions on what inventions to patent and facilitates with the negotiation of licensing agreements. As a result of this commercialisation role, the TTO has direct and frequent contact with industry. This contact can often spawn indirect effects such as student placement, donations to research and general support for the university. Although technically a step removed from the actual research, the TTO should have familiarity and good contacts with the expertise of faculty and should track potential industrial research opportunities. By conducting these activities, the TTO plays a central role in support of the HEI.

Among the respondent countries in the OECD survey (2003), TTOs in Germany, Japan and the Netherlands provide more IP services than TTOs in other countries. With the exception of Japan, TTOs do not usually license technologies from companies or other PROs. Overall, TTOs tend to be orientated towards managing patent and licensing, especially licensing contracts and research agreements with firms. However, in several countries such as Italy, Norway and Russia, TTOs appear less equipped to assess the patentability of new inventions. This may be attributable to the relative young age and inexperience of TTOs in these countries, in addition to weak incentives to patent.

Table 3.1: Managing Technology Transfer at MIT

MIT: Strategically choosing which technology transfer activities to conduct

Unlike many modern universities, MIT has no business incubation activities at all. The strategy of the technology licensing office (TLO) is to encourage as many invention disclosures as possible from faculty members by minimising the barriers to disclosure – currently, MIT discloses about 450 inventions per year. MIT's TLO then licenses these inventions as nonexclusive or exclusive licences to industry and local venture capital firms. Rather than getting involved in the complexities of spin-out formation, the TLO provides a shop window for industry to view its Intellectual Property and agrees as many licence deals as possible.

Licensing is less resource-intensive than spinning out new companies – both in terms of people and funding – and has a higher probability of getting technology to market. It is often the quickest and most successful way of transferring IP to industry, and has the advantage of using existing business expertise rather than building this from scratch.

Source: Adopted from Lambert (2003).

Anecdotal evidence from successful TTOs suggests that, as IP operations develop, TTOs expand their operations beyond patenting and licensing to develop contract/sponsored research and to provide technology consulting services, thus broadening their revenue base and generating more research for the institution. The appropriate approach will vary depending on the needs of local business, the mission of the university, and the focus of the local economy.

Management of Intellectual Property

A further role that TTOs are involved in is the management of intellectual property. The share of IP activities managed by the TTO for the host institution varies across OECD countries. 'The degree to which IP activities are concentrated in a TTO or administrative unit of the PRO tells much more about the TTO's formal organisation and capacity' (OECD, 2003: 41).

In Japan, 35% of PROs claim to manage 100% of IP for their institution, while 23% claim to manage between 25% and 75%. This reflects the use of separate or independent TTOs to manage IP for national universities. The figures for the Netherlands are 66% for universities. Only 38% of Norwegian PROs report that they manage 100% of the IP and nearly 30% report that IP is managed by other actors, namely the researcher. This reflects the fact that, in Norway, individual researchers own the IP; hence, the responsibilities for managing IP are more informal and diffuse. In Switzerland, 34% of university-based TTOs reported that they manage up to 100% of the IP of their institution. Thus, the extent to which IP is managed centrally by a TTO is contingent on IP policies and the underlying structure of the TTO.

Furthermore, the OECD survey found that, by and large, the Netherlands and Korea university-based TTOs are quite institutionalised in their organisation of IP. In Italy, in contrast, only 35.7% of universities rely on a dedicated TTO or licensing office, while 46.4% delegate such activities to other offices for which technology transfer is not the main mission.

Figure 3.1: TTOs Focusing on Technology Transfer, by Country

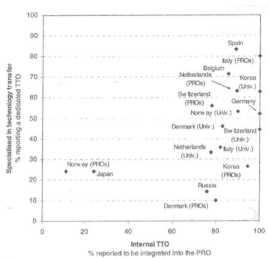

Some Empirical Evidence

Based on a survey and interviews with a number of TTO directors in the US and Japan, as well as AUTM data, Allan (2001) found that the main operations conducted by TTOs with respect to technology transfer included:

♦ Dealing with disclosure of inventions.

♦ Record keeping and management.

♦ Evaluation and marketing.

♦ Patent prosecution.

♦ Negotiation and drafting of license agreements.

♦ Management of active licenses.

Interestingly, many of the respondents noted that, while hundreds of thousands of dollars were often devoted to patenting and legal counsel costs, very little was set aside for marketing the resultant patents. Formal, proactive technology marketing programs were highly varied in the TTOs surveyed, with most doing very little active marketing. For most TTOs, available technologies were only listed selectively on their web sites. One respondent stated that, in his office, they literally wait for the phone to ring. In general, both large and small institutions market their intellectual properties primarily through Web-based posting services, and institutional home pages. Smaller TTOs typically also lack sufficient human and/or financial resources, which can impact the extent of such activities (Allan, 2001).

Network Development & Contact Mechanisms

An inherent part of TTO activities involves developing soft skills and maintaining contacts. Such relationship-building is an ongoing process that is crucial to the development of TTOs and to creating further opportunities for the commercialisation of research (Meyer, 2003). Research by Siegel *et al.* (2003) highlights the importance of such informal TTO activities. Siegel *et al.* (2003) interviewed 55 technology transfer stakeholders, asking questions in relation to importance of networks and relationships in HEI/business technology transfer. They found that 'relationship-building, personal relationships were mentioned more frequently than contractual contacts, emphasising the continual important informal aspect of communication and relationship building to the technology transfer process' (see **Table 3.2**).

Table 3.2: Aspects of Relationships/Networks in University-Industry Technology Transfer

Relationship/ Network	Managers/ Entrepreneurs	TTO Directors/ Administrators	University Scientists
Personal relationships	75%	66.7%	80%
TTO as a facilitator of relationships between scientists and firms	25%	75%	40%
Knowledge transfer from industry to faculty members	25%	20%	65%
Conference Technology Transfer meetings	35%	80%	15%
Contractual relationships	15%	6.7%	0%

Source: Siegel et al. *(2003).*

The values presented are the percentages of respondents who identified a particular item as an aspect of relationships/networks in HEI technology transfer. These findings indicate that a key role for TTOs should be the formation of networks, which, as all stakeholders emphasised, are important in technology transfer.

Research by Sanchez & Tejedor (1995) on the establishment of formal links between TTOs and business highlighted a number of key mechanisms, which include:

♦ Firm managers themselves search for university departments.

♦ Managers receive proposals from university departments.

♦ Managers seek support from TTO.

♦ Managers receive proposals from TTO.

A Typology of TTO Roles & Associated Activities

Based on the literature (Allan, 2001; Jones-Evans, 1999; Fassin, 2000: 38, Lambert, 2003: Scott *et al.*, 2001; Siegel *et al.*, 2003), as well as data obtained from various ILO websites, research evidence and primary interviews, we have developed a typology of TTO roles and associated activities. This model seeks to operationalise Fassin's (2001) three technology transfer roles (information broker, science marketer and catalyst for academic entrepreneurs) by relating them to activities conducted by TTOs (see **Table 3.3**).

Table 3.3: The Roles & Activities of TTOs

Role	Task	Detail
Information broker	Dissemination of information	Fielding inquiries from industry Development and distribution of brochures Promoting networking opportunities Acting as an information point Commercial advice
Science marketer	Marketing and promotional activities	Organising visits to laboratories Participating in conferences and presentations Participating in specialised technology fairs
Science marketer	Public Relations activities	Networking with professional associations Writing articles in periodicals and press Promoting special events Marketing the university and the industrial liaison function
Catalyst for academic entrepreneurs	Advice and negotiation of research contracts	Advising on intellectual property and patent issues Negotiating agreements Licensing Defining strategy for technology transfer
Information broker / Science marketer / Catalyst for academic entrepreneurs	Active management and valorisation of the HEI potential (technology transfer)	IP management: identify/evaluate/protect Searching for industrial partners Searching for commercial partners Searching for financial partners (venture capital funds, business angels) Initiating spin-off companies, helping with business plans
Information broker	Collection of information	Building information systems such as internet home pages, databases for partner search Maintaining of directories of research activities and expertise
Information broker / Science marketer / Catalyst for academic entrepreneurs	Co-ordination	Of the university research park Of the university incubator centre Of the university seed capital fund Work placements/education/distance learning

RESPONSIBILITIES OF TTOS

The ultimate responsibility of TTOs is 'guardian of the university's intellectual property' (Fassin, 2000: 37). Yet this is narrow in its depiction, as often TTO's responsibilities extend beyond the mere protection of formal IP. As noted by the OECD (2003: 59), 'TTOs do far more than simply ensure the protection of patentable inventions. They are often involved in protecting and exploiting innovations in a number of technological fields'. TTOs often receive invention disclosures and are responsible for deciding which innovations and creative works to protect and how to do so. They may also be involved in negotiating contracts that stipulate how IP is to be used or revealed. In this role, the responsibility of the TTO is not to get the patent but rather *to close the deal*. The TTO's priority, therefore, is getting the technologies into the marketplace for the public good (Meyer, 2003: 11). Secondary responsibilities include promoting technological diffusion and securing additional research funding for HEIs *via* royalties, licensing fees, and sponsored research agreements.

The main responsibility of TTOs – to assess and protect IP and make it available for public use (OECD, 2003) – is evident from the sample mission statements in **Table 3.5**, which depict the key responsibilities and roles of a number of TTOs.

TTO Responsibilities to Multiple Stakeholders

TTOs are subject to a number of competing demands and have to meet the needs of various stakeholders, including faculty, administration, commercial and state priorities (Graff *et al.*, 2001). These multiple responsibilities are particularly evident in some activities – for example, the provision of legal and intellectual property management services to researchers and the collection of licensing royalty revenues for the HEI from industry. The revenues generated by licensing, while still only a minor percentage of HEIs' operating budgets, have grown substantially. Beyond their monetary value, the growth of such royalty revenues serves to demonstrate the success of TTOs in diffusing the fruit of the HEI's research.

In a study undertaken to review the economic effects of the Bayh-Dole Act, Jensen & Thursby (2001) surveyed 62 research universities concerning their technology transfer activities for the years 1991-1995. TTOs, faculty, and the university administrators were all interviewed about their objectives for technology transfer. The different parties' responses demonstrate the differences that exist within the HEIs over technology transfer (see **Table 3.5**).

Table 3.4: The Responsibilities & Roles of TTOs:
Sample Mission Statements

University	TTO Mission
Warwick University	The University of Warwick is committed to maximising the commercial application of its research to benefit the regional and national economy. Warwick Ventures spearheads this mission through patenting, licensing and creating spin-off companies based on selected research innovations.
Science & Technology Ventures, Columbia University	The Science & Technology Ventures mission is to evaluate, protect, and license Columbia intellectual property, increase private sector funding for research and development, encourage technology transfer, and distribute income from these activities among Columbia entities and faculty.
Florida Office of Research Technology Transfer	Encourage and assist technology development at Florida State University and facilitate the transfer of intellectual property to business and industry to provide benefits to the university, the economy and to improve the overall quality of life.
Northwestern University	The mission of the Technology Transfer Program is to facilitate transfer of the University's innovative technologies for public use and benefit through licensing and other commercial agreements, to support and to assist with the formation of spin-off companies, while adhering to the basic academic principles; to aid faculty, staff and students in matters relating to the protection and marketing of their intellectual property; and to provide an additional source of unrestricted income to support research and education at Northwestern.
Harvard Office for Trademark & Technology Licensing	To bring University-generated intellectual property into public use as rapidly as possible while protecting academic freedoms and generating a financial return to the University, inventors and their departments. To serve as a resource to faculty and staff on interactions with industry. To protect against unauthorized third-party use of the University's various trademarks worldwide and to license their use on approved merchandise, generating income for support of undergraduate financial aid.

**Table 3.5: Divergent Priorities: Ranking the Importance of
University Technology Transfer Outcomes by TTO
Officers, Administrators & Faculty**

Outcome	Priority Ranking by		
	Technology Transfer Officers	**University Administration**	**University Faculty Members**
Revenue	1	1	2
Inventions commercialised	2	3	3
Licenses executed	3	4	5
Sponsored research	4	2	1
Patents granted	5	5	4

Source: Jensen & Thursby (2001).

Administrators tend to consider technology-associated revenues most important, while the faculty see the ability to attract research sponsorship as paramount. TTOs often operate under somewhat conflicting mandates from their administration and faculty and emphasise the more immediate and tangible outcomes of executed licenses and commercialised inventions.

Priorities of TTO Staff

Beyond what the various parties report to be priorities for technology transfer, actual investments of time by TTO staff in particular activities reveal how TTO staff respond to (and perhaps balance) the conflicting demands placed on them (Castillo *et al.*, 2000). TTO officials undertake multiple tasks: they scout and assess inventions at their institutions, make patent applications, market patent rights to possible buyers, mediate contacts between their institution's personnel and outside investors, and monitor and enforce licensing and research contracts.

The time spent on a successful invention by a TTO is typically split 50-50 between efforts expended prior to, and after, patenting. Pre-patenting activities involve solicitation of invention disclosures from faculty, evaluation of inventions, and assessment of their economic potential. Patent preparation involves, on average, only about 10% of the total time and efforts of TTO staff. Commercialisation efforts, including the negotiation of a license, involve about 25% to 30%, and follow-up activities of monitoring and enforcement involve about 10% of TTOs' expended time (see **Table 3.6**).

Prior to the mid-90s, private HEIs appeared to put more effort into securing earnings from technology transfer. Private HEIs emphasised market-related activities (evaluating inventions and assessing markets) and spent relatively more time enforcing and monitoring contracts.

Table 3.6: Revealed Priorities: Percentage of Time on the Job that Public & Private University TTO Officers Spend on Various Activities, 1999

Activity	Percentage Time Spent by Universities		
	Public	Private	Both
Soliciting ideas	11.3	7.8	10.0
Evaluating inventions	14.1	16.6	15.0
Assessing markets	12.5	13.6	12.9
Referring inventors	3.8	2.4	3.3
Preparing patent applications	8.6	10.1	9.1
Drafting licenses	24.3	24.6	24.4
Enforcing patents	2.8	5.0	3.6
Monitoring contracts	7.2	11.5	8.7
Others	15.8	10.1	13.9

Source: Castillo et al. (2000).

Identifying TTO Stakeholder Characteristics

It is critical to appreciate the various actions, motives, and organisational cultures of university/industry technology transfer stakeholders. In order to understand the responsibility of TTOs in relation to the key demands of alternative stakeholders, it is useful to draw again on Siegel's *et al.* (2003) study (see **Table 3.7**).

Siegel *et al.* (2003) assert that a primary motive of university scientists is recognition within the scientific community, which emanates from publications in top-tier journals, presentations at prestigious conferences, and research grants. This is an especially strong motive for untenured faculty members. Other possible motives include financial gain and a desire to secure additional funding for graduate assistants, post-doctoral fellows, and laboratory equipment/facilities. The norms, standards, and values of scientists reflect an organisational culture that values creativity, innovation, and, especially, an individual's contribution to advances in knowledge (basic research).

Table 3.7: Characteristics of University Technology Transfer Stakeholders

Stakeholder	Actions	Primary Motive	Secondary Motive	Organisational Culture
University scientist	Discovery of new knowledge	Recognition within the scientific community	Financial gain and a desire to secure additional research funding	Scientific
TTO	Works with faculty and firms/ entrepreneurs to structure deal	Protect and market the university's IP	Facilitate technological diffusion and secure additional research funding	Bureaucratic
Firm/ entrepreneur	Commercialises new technology	Financial gain	Maintain control of proprietary technologies	Entrepreneurial

The TTO must work with scientists and managers or entrepreneurs to structure a deal. Firms and entrepreneurs, on the other hand, seek to commercialise university-based technologies for financial gain. They are also concerned about maintaining proprietary control over these technologies, which potentially can be achieved through an exclusive worldwide license. The entrepreneurial organisational culture of most firms (especially start-up companies and high-tech firms) rewards timeliness, speed, and flexibility.

In summary, there are palpable differences in the motives, incentives, and organisational cultures of university-industry technology transfer stakeholders that potentially could impede technological diffusion (as illustrated by research by Castillo, 2000; Scott *et al.*, 2001 and Siegel *et al.*, 2003). One of the main responsibilities of TTOs is managing expectations, negotiating deals and communicating in a manner that potentially diffuses or helps to resolve these differences. Such connections and understanding can also be directed by government policy – for example, Science Foundation Ireland awards, which foster the natural connections between researchers in an academic environment with their counterparts in industry. TTOs also face frequent pressure from funding agencies to formalise practices with regard to such issues.

TTO STAFFING & DEVELOPING EXPERTISE

Critical to developing HEI third-stream activities is investment in staff and obtaining the optimal mix in terms of experience and skills sets. The extent of developing such expertise in HEIs is subject to resource constraints and the strategic priority that HEI administrators place on third-stream activities.

Levels of Staffing

Surveys of OECD countries show that TTOs are recent and generally have fewer than five full-time staff (OECD, 2003: 12). Staff at TTOs are at the forefront of technology transfer and, as already outlined, are required to liaise with scientists and patent attorneys as well as funding bodies. The results from the OECD show that, for most countries, the staff numbers at TTOs are relatively low but are growing. In Norway, for example, only 1 out of 5 of survey respondents has more than one full time equivalent (FTE) dedicated to technology transfer issues. In the US, the number of TTO staff is somewhat larger, with a mean of 3.3 staff devoted to licensing issues (**Table 3.8**).

Table 3.8: Employment at TTOs at US Universities
(full time equivalents), 2000

	Licensing staff	Other staff
Total	562.5	586.5
Mean	3.3	3.5
Median	2.0	1.8
No of responding universities	168	168

Source: US Technology Administration, Department of Commerce, based on AUTM 2000 data.

In an Irish context, not all HEIs have clearly identified the role and function of IP management or have created specific positions to manage it. In these institutions, approximately 62 people (22 full-time equivalents) are regarded as being involved in commercialisation activities – providing an average of 0.96 FTE staff per research institution. This compares with a median of 2.2 FTE licensing staff and 1.8 FTE other staff at the TTO offices in the US. Resources at TTO offices at HEIs in Ireland are thus deemed inadequate to ensure successful commercialisation of research (Forfás, 2004a).

An examination of the cost of technology transfer management in the US suggests that institutions with an annual R&D expenditure of less than €40 million cannot recoup their technology transfer commercialisation expenditure. In the UK, the *National Health Service Guide* states that the cost of running a technology transfer unit employing two full-time staff, plus clerical support, could amount to €250,000 *per annum* including overheads, plus an additional €100,000 in patenting costs (including patent attorneys' fees), giving a total of €350,000. Based on a return of 2.5%, such an overhead requires an annual research investment of at least €14m to €20m.

Technology Transfer Skills

In the US, the land grant universities (those that have been granted land upon which to build, in return for delivering services back to the local community) have had the application of knowledge at the heart of their mission from their inception. This has helped in the development of highly-integrated programmes, such as North Carolina State University's TEC program, as well as extensive skills and experience in managing technology transfer. This contrasts with the situation in many EU countries, where commercialisation activities are not historically supported and grounded (PACEC, 2003).

Technology transfer is people-intensive and requires a wide and specialist set of skills. Many HEIs face problems in building professional offices on their own. Protecting and managing IP requires specific legal knowledge. Licensing needs a combination of market awareness, subject-specific knowledge, marketing and negotiating skills. Spin-out creation requires entrepreneurship skills, links with business angels and venture capitalists, business planning, management and company formation expertise. These skills are difficult to find in a small group of people and are expensive to buy in. Only a small number of HEIs have built up licensing expertise (OECD, 2002; 2003).

New Skill Sets

The massive expansion in the roles, activities and responsibilities of TTOs has meant that an increasingly more diverse and advanced skill set is required of personnel employed in this arena. These demands are further complicated by multiple stakeholder influence.

The new range of skills required by TTOs, as identified by Shattock (2001), includes:

♦ The ability to build networks.
♦ A capacity for brokerage.
♦ A wider vision about the university and the economy.
♦ Ability to marrying market niches and gaps.

◆ Strategic skills in identifying university research strengths.

◆ Legal and intellectual property skills.

◆ Skills in company formation.

Personnel Profile

The most successful TTOs in the US – such as those at MIT and Columbia – place strong emphasis on recruiting staff with substantial industry experience, and find it difficult to teach the negotiation and deal-making skills learnt in industry to new staff. The tendency in Europe is that TTOs are staffed by academics or HEI administrators. This can create barriers to the negotiation of contracts, as business generally finds it easier to negotiate with individuals who have more of a commercial background. Yet a major restriction in addressing this deficiency is the limited salary to attract experienced entrepreneurs or industry executives into TTOs.

The Lambert Report (2003) noted that UK advertisements still focus on subject background, rather than functional experience. Allan's (2001) study of best practice found that TTOs noted difficulties in training faculty in patenting and the innovation process. Of the institutions that were studied, TTOs with longstanding operations seemed to draw more from years of precedence and reputation to maintain and enhance the technology transfer process (Allan, 2001). EMIS (1995) noted, however, that often researchers will have industrial experience 'some of the most successful technology transfer programmes are invariably initiated and taken care of by senior researchers who both excel academically in their specialist discipline and have solid industrial experience or have held senior positions in related industries'. Having capable staff is crucial as 'well-trained staff at TTOs are not only essential to the efficiency of technology transfer but can also help limit conflicts of interests with researchers' (OECD, 2003: 46).

Training needs to be both formal and informal, and should address the skill deficits in the following areas:

◆ Appropriate research protocols.

◆ Logging and recording of research findings.

◆ The patent process, patent protection, and patent law.

◆ The costs of commercialisation.

Meyer (2003: 12) provides a useful list of the essential characteristics of staff employed by a TTO (see **Table 3.9**).

Table 3.9: Essential Characteristics of Professional Staff Employed by TTOs

- Have the passion to serve the faculty and industry and to find a way to close the deals in a prompt manner, consistent with university policy.

- Be an entrepreneur with education and experience in an appropriate scientific, engineering, technology background to complement the research strengths of the campus.

- Have the ability and enthusiasm to market technologies.

- Be knowledgeable in intellectual property and contract law.

- Be a good negotiator. This is a person who is well-informed on the terms of comparable deals and willing to provide the data, reasonable in offering and accepting terms, and displays an attitude of flexibility in negotiations to get a win-win outcome in a reasonable time.

- Have excellent communication skills.

Source: Meyer (2003: 12).

Functions of TTO Staff

Allan's (2001) best practice study found that most TTOs stress the importance of informal relations and relationship-building with academics. This may suggest why, in many cases, it is recommended that staff hired for TTOs have industry experience and, hence, an understanding of industry as well as a network of contacts and linkages (OECD, 2003). A study by Balthasar *et al.* (2000) found that 'successful institutions at the interface between science and industry do not consider themselves to be an institution for transfer but a network manager'. The functions and activities carried out by Technology Transfer Managers/Directors are multiple, as illustrated in **Table 3.10** below.[13]

---

[13]   Based on EIMS (1995), Scott *et al.* (2001) and sample job descriptions from Georgetown University provided by Dr. Martin Mullins.

Table 3.10: Functions/Activities of Technology Transfer Managers/Directors

◆ Develop and implement an intellectual property and technology commercialisation strategy.

◆ Contribute to the development of institutional policies relating to licensing of intellectual property.

◆ Responsible for institutional patent prosecution and management strategy.

◆ Screen technological change and market evolutions.

◆ Identify and evaluate inventions generated by researchers.

◆ Adopt technology transfer options to meet continuously-changing demand.

◆ Search for new clients.

◆ Define collaborative projects.

◆ Ensure professional project management.

◆ Manage patent filing and patent prosecution.

◆ Negotiate confidentiality, option, licensing, sponsored research and collaboration agreements.

◆ Managing working relationships with industrial clients and researchers.

◆ Facilitate new company formation.

◆ Counsel faculty, staff and students in matters relating to intellectual property, licensing and conflicts of interest.

◆ Deliver outreach programs to faculty, staff and students.

KEY ISSUES FOR HEIS

The implications for HEIs are clear. In order to develop technology transfer initiatives, HEI leaders and administrators have to have a clear strategic focus and plan for such activities. In practice, this means developing a strategic plan that is cognisant of the current status of such activities within their own institutions, as well as further aspirations that are aligned to different stakeholder requirements. Consequently, by clearly defining roles, activities and responsibilities of TTOs, Directors of such offices can manage technology transfer activities effectively and become a guardian and champion of the HEI's intellectual property.

Specifically, based on this chapter, some key issues can be extrapolated:

♦ Each TTO should have a clear mission statement and strategic plan, where key roles, activities and responsibilities are clearly defined.

♦ Technology transfer activities involve a mix of roles, including information broker, science marketer, and a catalyst for academic entrepreneurs.

♦ One of the key responsibilities of a TTO is to be the guardian of the university's intellectual property.

♦ Stakeholders of TTOs have different motives and incentives. Consequently, one of the main responsibilities of ILOs is managing expectations and communicating with the stakeholders in a manner that potentially diffuses, or helps to resolve, these differences.

♦ Soft skills in technology transfer are crucial to creating further opportunities for the commercialisation of research.

♦ TTOs place significant efforts in training faculty in patenting and in the innovation process.

♦ The rapid growth of TTOs roles and activities requires them to develop a new range of skills, including:

◊ the ability to build networks;

◊ a capacity for brokerage;

◊ a wider vision about the university and the economy;

◊ ability to marrying market niches and gaps;

◊ strategic skills in identifying university research strengths;

◊ legal and intellectual property skills;

◊ skills in company formation.

♦ The most successful TTOs place a strong emphasis on recruiting staff with substantial industry experience.

4

MECHANISMS FOR TECHNOLOGY TRANSFER: PATENTING, LICENSING & COMPANY FORMATION

The exploitation of knowledge and commercialisation of research must become embedded in the culture and infrastructure of the higher education system. This requires ... new campus company start-ups, a pro-innovation culture of intellectual property protection and exploitation... and greater links between higher education institutions and private enterprise.
Enterprise Strategy Group (2004)

INTRODUCTION

In order for the benefits of HEI research to be expressed in the economy, the HEI research system has to be connected with the economy. Much of the economic literature assumes that such connections come about as a result of 'spillovers' – side effects or 'externalities' of public research. This idea has been criticised by a number of authors for tending to 'black box' the specific mechanisms through which knowledge can flow (Scott *et al.*, 2001). In this chapter, we explore such mechanisms for technology transfer. In the past, technology transfer has often been informal or occurred in an unplanned, *ad hoc* and fortunate manner. While this sort of transfer is an inherent part of the diffusion and exploitation of research results, more formalised mechanisms have been set up under the auspices of TTOs or ILOs at various HEIs.

We begin the chapter by examining some of the theoretical debates about technology transfer and research commercialisation. We then focus on the commercialisation of research, and technology transfer in the form of patenting, licensing and company formation. This consideration is critical as 'the creation of intellectual property and its commercialisation has become a vital competitive factor in the modern industrial economy' (ICT Ireland, 2004a). We conclude by examining softer mechanisms for technology transfer, such as collaborative research agreements, education and consulting activities.

SOME THEORETICAL DEBATES: RESEARCH COMMERCIALISATION & TECHNOLOGY TRANSFER

Harmon *et al.* (1997) classified research commercialisation and technology transfer literature into two groups, a rational decision-making perspective and a relationship perspective.

Rational Decision-making Perspective

This perspective views technology transfer as a process that can, and indeed should, be planned. Inventors and future users of the technology are seen to function independently, without co-ordinating their efforts until the first negotiations regarding a specific technology. The majority of the studies following this line of thought focus on the process of technology transfer from the research centre to industry. The major focus is to identify the most efficient methods of administering and facilitating the technology transfer process and the organisational forms that facilitate this transfer. This linear model of the innovation process is based on factors such as basic research, applied research, prototype development, market research, product development, marketing and selling.

This is the model that has had the greatest influence on public policy in most countries. Interventions are made at different and specific stages by strengthening public infrastructure, and improving incentives to the private sector, which was then expected to transform technology, patents and systems into new products and processes. This group of studies includes several models. The first model, a 'Brownian model' developed by Padmanabhan & Souder (1994), views successful technology transfer as a process of managing a portfolio of interacting facilitators and barriers in such a way as to bring about successful technology transfer. A second model, developed by Goldhor & Lund (1983), envisages technology transfer as a bridge-crossing process facilitated by a transfer agent, who is seen as the cornerstone to the process.

The Relationship Perspective

The second major group of studies, reviewed and categorised by Harmon *et al.* (1997), takes a different perspective on the technology transfer process, emphasising the importance of relationships. This group of studies is primarily made up of non-linear models that emphasise multi-directional linkages, and interdependency between hard technology and the softer issues of people management and information flows (Mitra & Formica, 1997). A number of perspectives are found in this group of studies:

♦ A **communications** perspective, which states that a successful transfer depends on the effectiveness of information flows between a set of individuals or organisations within a complex network of communication paths.

♦ The **innovation** perspective, which argues that policy-makers should not be as concerned with processes for transferring technology after it has been developed, but should focus instead on developing transferable technologies in the first place.

♦ An **alliance** perspective, which argues that barriers can be reduced by alliance building between inventor and lead users, thus facilitating the identification of marketplace demands and requirements. This alliance then develops into a relationship with the lead user facilitating the technology transfer at a later stage.

♦ A **co-operation** perspective, which studies the processes of co-operation between the parties involved that make the transfer easier. Among the facilitating processes identified in these studies are: open communication, mutual interdependence, respect, trust and willingness to compromise.

The modern view of research commercialisation and technology transfer tends to be grouped within the latter view. Increasingly, there is awareness of the multitude of factors that can impact on the technology process or failure of research to progress on this route (see Bozeman, 2000; Friedman & Silberman, 2003; Scott *et al.*, 2001; Siegel *et al.*, 2003).

HEI – INDUSTRY COLLABORATION

An inherent underpinning to the existence of TTOs is the concept of collaboration. Collaboration can take many forms, ranging from arms-length relationships to intense research initiatives sharing funds and ideas. Essential to the concept is the idea that 'any joint activity should be mutually beneficial' (Fraiman, 2002).

There are a range of activities that shape and influence the relationship between universities and the rest of society. Types of collaboration cited in the literature include student internships, jointly-sponsored symposia, joint research projects, periodic meetings, guest speakers; executive development – for example, MBAs, roundtables, joint curricula development, outreach programmes, and alumni associations/bodies.[14] Much collaborative activity is based on informal networks and contacts. Research relationships, therefore, are only a subset of the many different interactions between HEIs and commercial entities.

TTOs & University-Industry Collaboration

Initially, TTOs were set up in response to the desire to commercialise research for the public good. Graff *et al.* (2001: 14) note that two other key objectives have spawned from this: the provision of legal and intellectual property management services to HEI researchers and the collection of licensing revenues for the HEI. The roles, activities and responsibilities of TTOs were discussed in **Chapter 3**. The guiding principle for TTOs should be that they have a policy on technology transfer that is consistent with the broader mission of the university (OECD, 2003). Technology transfer usually is not a large revenue generator. MIT, Stanford and Yale all state that their main reason for engaging in technology transfer is to improve the public good – that is, to create the greatest possible economic and social benefits from their research, whether they accrue to the university or not (Bok, 2003).

Table 4.1: Encouraging University-Industry Collaboration: Matching Services

Matching Service: Finland

Sitra, a company established by Finnish statute, conducts and commissions research, provides loans and other financing, awards grants, offers surety and guarantees, and participates in development projects, as well as owning stocks and shares in companies. Seed finance and commercialisation of technologies are two of its operational areas.

In 1996, Sitra launched its matching service as a meeting place for private investors and entrepreneurs. The objective is to link investors, venture capitalists and entrepreneurs, to improve the flow of investment capital and management expertise into start-up and growth companies and to offer better opportunities for syndicate investments. In addition, Sitra operates a commercial mechanism - INNOTULI - which is a phased exploitation fund covering feasibility, proof of concept and venture capital stages.

---

14   This list of potential collaborative activities may serve as a useful template, although it is crucial to recognise that different disciplinary areas will have different objectives and, therefore, necessarily will have preference for the collaborative ventures that are most beneficial to them (Fraiman, 2002).

Siegel *et al.* (2003) note that the key objective for most TTOs is facilitating technological diffusion through the licensing to industry of inventions or intellectual property resulting from university research. Evidently, there are a diversity of models and channels for transferring knowledge and technology from the public to private sectors, including:

♦ University research sponsored by companies.

♦ Faculty consulting.

♦ Licensing of university-owned intellectual property to existing companies.

♦ University support for start-up companies in the form of loans, grants, and equity ownership.

♦ 'Mega agreements' between individual companies and universities that cover a range of interactions.

♦ Research centres and other government-supported efforts to encourage university-industry collaboration.

♦ Industry consortia to support university research.

The most common of these tends to be collaborating with business on research projects and agreeing at the outset exploitation rights on any IP created, and making deals with companies to exploit IP already developed in university research. The first approach depends on experienced negotiators from both parties agreeing terms and conditions for IP within an adequate framework. The second also requires dedicated expertise in licensing, spinout creation, venture capital, market research, marketing and IP management. Scott *et al.* (2001) also provide a summary of the key technology transfer and research commercialisation methods used by HEIs, depicted in **Table 4.2**.

The first three of these mechanisms are more collaborative in nature, while the others involve licensing already completed research for commercialisation purposes. Licensing and exploitation of intellectual property are the more specific forms of technology transfer activity arising out of research. Patenting contributes to the effective use of technology, in that a company is more likely to invest in new technology if it is protected by a patent. Usually, it is the licensing of university technology that is referred to as *technology transfer*; but true technology transfer is accomplished through all of these mechanisms, as well as through the use of faculty consultants and extension services.

Table 4.2: TTO Research Commercialisation/Technology Transfer Mechanisms

Commercialisation / Transfer Mechanism	Description
Sponsored research	The most frequent form of research relationship, which involves companies directly funding university research.
Collaborative research	University-industry research partnerships that can be encouraged through partial government funding.
Technology licensing	Licensing of university patents (usually stemming from federally-funded research) to companies for commercialisation.
Start-up companies	Obtain licensing agreements to access university technologies.
Spin-offs	Includes among its founding members a person affiliated with the university
Exchange of research materials	Used to expedite the performance of research and accomplished through material transfer agreements.

Company formation is seen by many as being the most direct manifestation of technology transfer. Chalmers University in Gothenberg, for example has developed a strong tradition in promoting academic entrepreneurship, generating approximately 40 to 50 companies per year and several thousand jobs.

Materials transfers are strongly related to research commercialisation activities. Materials transfer agreements (MTAs) are made on the basis that the technical equipment available in universities and public research laboratories can also be used by existing local business to solve production process problems and to supplement their commercial advantage.

Consulting can be the most versatile and cost-effective means of linking industry with the university and public research sector, as it is a relatively inexpensive, rapid and selective means of transferring information with few institutional tensions and it rarely involves extensive demands on research personnel or material resources (Stankiewicz, 1986).

COMMERCIALISATION OF RESEARCH:
SOME EMPIRICAL EVIDENCE

Commercialisation can be defined as a process by which research outputs are converted to commercial usage or ownership (Forfás, 2004a: 15). Howells & MacKinlay (1999) conducted surveys across 10 universities in Europe to ascertain the criteria for effective technology transfer. They found that the effective exploitation and management of intellectual property rights and technology transfer within a university was based on a number of activities that run in parallel.

Table 4.3: Success Factors for Commercialisation &
Technology Transfer Activities

♦ Effective monitoring of research and inventive activity that is being generated by the university

♦ Accurate identification and selection of research inventions that are seen as valuable and worth protecting in terms of generation of future income

♦ Comprehensive technology auditing to pick up non-disclosure of intellectual property being generated by academics and owned by the university;

♦ Flexibility in decision-making as different technologies and circumstances favour different routes to commercialisation and timing;

♦ Take-up and negotiation of the selected research and inventions with the research team for the protection and defence of the research involving the establishment of appropriate incentive and exploitation schemes.

♦ Selection and establishment of the appropriate legal IPR defence mechanism for the research and inventions

♦ Appropriate decision on the long term IPR exploitation and development route of research and inventions.

Source: Howells & MacKinlay (1999).

Commercialisation Strategies

The OECD (2003) has noted the importance of making the appropriate decision as to the exploitation route and the technology transfer mechanism chosen (see also Friedman & Silberman, 2003). The suitability of the mechanism chosen will be contingent on the HEI's institutional context and commercialisation objectives. Howells & MacKinlay (1999) found that much of the commercialisation practice in European HEIs is not about doing what is 'best', but rather involves taking what is the 'least cost' route. Achieving the 'best', in the sense of leveraging the maximum long-term financial and commercial return for the HEI, was considered a rare event for most HEIs and not unexpectedly is reserved for a tiny number of elite projects considered likely to do well commercially. This echoes the focus on 'getting the deal done' as discussed in **Chapter 3**. For the more exceptional projects, the most likely preferred exploitation route will be *via* a company start-up, if the academic inventors are willing to become entrepreneurs.

Howells & MacKinlay (1999) also reported that academic respondents referred to a number of factors that facilitated and encouraged them to commercialise their research results. These included:

♦ Clarity in the frameworks and procedures of exploitation and commercialisation policy.

♦ Transparency in decision-making concerning exploitation and commercialisation activity.

♦ Good information and contact provision to academic personnel regarding intellectual property and commercialisation procedures.

♦ Streamlined decision-making procedures.

In a number of countries, government agencies have been active in supporting commercialisation strategies at HEIs. Enterprise Ireland, for example, offers a number of support programmes that provide assistance to college researchers in technology commercialisation (Forfás, 2004a: 16). Further, through its Commercialisation Fund, Enterprise Ireland provides funding for research commercialisation, as follows:

♦ Proof of concept: €50,000 to €90,000.

♦ Technology development: €300,000 to €450,000.

♦ Commercialisation of R&D: 50% of eligible expenditure up to €38,000.

In addition, Enterprise Ireland has a range of other supports that could be helpful to assisting R&D and patent commercialisation – for example, seed and equity funding for start-up companies, the Incubator Centre Initiative to support space for spin-off enterprisesand Mentor programmes.

As well as clear policies and agency support, further contingencies that may influence commercialisation activities and effectiveness include the characteristics of the transfer agent as discussed below (potential barriers and stimulants to technology transfer are discussed in-depth in **Chapter 5**).

Characteristics of the Transfer Agent

Many of the difficulties that may arise in commercialisation activities will be related to cultural resistance by researchers or sections of institutions as to the validity of such activity and its impact on research activities. Much of this can be attributable to concerns of over academic freedom, as researchers have traditionally been motivated to publish research findings as soon as possible for reasons relating to status and career development.

In a study of 1,000 academic researchers from the top 100 US research universities, Rahm (1994) found that 25% of researchers were 'university-bound' and 75% were 'spanning researchers'. The spanning researchers tended to initiate communications with industry and were more likely to have informal links with industry. Three-quarters of the spanning researchers engaged in consulting (against of the 26% university bound) and 80% of spanning researchers had students in industry with whom they had regular contact (18% of university bound). Similarly, 'spanners' were more likely to participate in research consortia, incubators and co-operative R&D.

The results of a survey by Fischer (1994) found that technology transfers between suppliers and receivers that have past business relationships are generally more successful than transfers to unknown sources, which are sometimes opportunistic in nature. Rahm *et al.* (1998) examined how the impact of characteristics and the composition of university and public research laboratories impacted on their participation in technology transfer. From a survey of 665 researchers, they found that those involved in basic research were less likely to engage in technology transfer activities. Further, the strongest predictor of technology transfer was seen to be the presence of diversity in research missions. Those laboratories that were narrowly focused, regardless of the nature of their focus, were less likely to be engaged in technology transfer. Bozeman & Coker (1992) found that three different types of effectiveness related to the attributes of the transfer agent, but in different ways. The number of licenses related chiefly to the size of the laboratory. Getting technologies 'out the door' was best explained in terms of the missions of the laboratories and the composition of R&D, while research diversity and degree of commercial orientation of the lab best explained market impact, measured in terms of commercialised technology.

Commercialisation Success in the UK

In the UK, the so-called 'golden circle' of Oxford, Cambridge and Imperial College London has been the most successful in commercialising their research. Oxford has long-standing contacts with industry and provides incentives to enterprising academics by ring-fencing their equity stakes against too much dilution. Cambridge, traditionally Britain's premier institution for scientific research, has an alliance with MIT, which has opened doors to lucrative commercial contracts. It is also the geographical and spiritual centre of 'Silicon Fen', Britain's high-technology hub. Imperial College has hired former GlaxoSmithKline chairman, Sir Richard Sykes, as its rector. Sykes is a passionate advocate of better links between universities and industry and, as well as tapping his old company for funds, has spoken of the commercial value of the Imperial brand. Other universities to have floated spin-out companies include Newcastle, Durham, Edinburgh, Glasgow and, previously, the University of Manchester Institute of Science & Technology (UMIST).

TECHNOLOGY TRANSFER

Technology transfer refers to the process whereby invention or intellectual property from academic research is licensed or conveyed through use rights to a for-profit entity and eventually commercialised (Friedman & Silberman, 2003: 18). In other words, it is the means by which innovative ideas move from the laboratory to the marketplace (Forfás, 2004). According to the AUTM, the phrase 'technology transfer' can be used very broadly to define the movement of ideas, tools, and people among institutions of higher learning, the commercial sector and the public sector (AUTM, 2002) (for a discussion of definitional complexities, see Bozeman, 2000). Dean & LeMaster (1995) note that technology transfer includes 'the purpose, application, and justification of the technology. It is what to employ, how, when, and why'.

Generally, technology transfer takes place along a research continuum ranging from transfer of materials agreements, to initial patent applications, assistance in identifying/evaluating new disclosures, to identifying and negotiating license agreements, to the most direct manifestation of technology transfer in terms of managing company creation. The AUTM has highlighted the economic impact of successful technology transfer on the US economy since 1993. For example, in 2002, running royalties for higher education institutions on product sales were $1,005 billion, an 18.9% increase over the fiscal year 2001. Friedman & Silberman (2003) capture the process of technology transfer as outlined in **Figure 4.1**.

Figure 4.1: The Process of University Technology Transfer

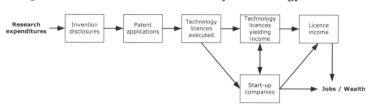

Source: Friedman & Silberman (2003: 18).

Technology Transfer Process

Technology transfer begins with industry and HEI interaction. Some argue that this normally follows a predictable sequence (Van Dierdonck & Debackere, 1988).

The sequencing of events in terms of university technology transfer is usually that a scientific discovery is made. The scientist then files an invention disclosure with the TTO. Once formally disclosed, the TTO must simultaneously evaluate the commercial potential of the technology and decide whether to patent the invention, as well as deciding the geographical extent of the patent protection. This decision often poses a dilemma for many TTOs, as they have limited resources for filing patents. If the patent is awarded, the TTO will often attempt to market the technology. Faculty members are frequently involved in the marketing phase, as they are often in a good position to identify potential licensees and because their technical expertise often makes them a neutral partner for companies that wish to commercialise the technology (Siegel *et al.*, 2003). The final stage involves the negotiation of a licensing agreement. These agreements can include benefits to the university such as royalties, 'follow on' sponsored research agreements, or an equity stake in a new venture based on a licensed technology. Involvement does not necessarily end with a licensing agreement; instead, it is quite common for TTOs to devote substantial resources to the maintenance and recognition of licensing agreements

The technology transfer process, therefore, typically includes a set of components: starting with investment in R&D; the actual R&D performance; the decision how to handle intellectual property; building a prototype to demonstrate the technology; the further development needed for commercialisation; and finally resulting in the successful introduction of a product or service on the market. This depiction is based on the premise of an ideal sequencing of technology development as shown in **Table 4.4**.

Table 4.4: Technology Transfer: Stage Activity Outcomes

- Basic research patent.
- Proof of concept/invention invention (functional).
- Early-stage technology development business validation.
- Product development innovation: new firm or programme.
- Production/marketing viable business.

Source: NIST (2002).

More Complex Technology Transfer

Often the process is more complex than depicted in **Figure 4.1.** Evaluation of the invention disclosure, for example, can involve multiple actors, including state bodies or the inventors themselves. Furthermore, **Figure 4.1** places too much emphasis on patenting, when it can be the case that technology is licensed prior to a patent being issued (Scott *et al.*, 2001; Siegel *et al.*, 2003).

Overall technology transfer activities are manifested as eliciting and processing invention disclosures, licensing university-created knowledge, seeking additional sponsorship of R&D projects or a combination of these three. Crucially, the extent of, and mechanisms for, technology transfer will be *shaped by* the resources, reporting relationships and incentives of technology transfer offices (Bercovitz *et al.*, 2001).

The complex nature of technology transfer is exemplified when one considers research in this area. Drawing largely on AUTM data, various authors (Thursby & Kemp, 2002; Rogers *et al.*, 2000; Politz *et al.*, 2000; Siegel *et al.*, 2003; Thursby *et al.*, 2001; Carlsson & Fridh, 2002; Friedman & Silberman, 2003) have used wide-ranging variables to demonstrate technology transfer effectiveness. Results offer insightful, yet often contradictory, conclusions. For example, research by Rogers *et al.* (2000) stresses the significance of faculty quality, while Thursby *et al.* (2001) finds that this is not a significant variable in explaining variances in efficiency between TTOs. **Table 4.5** summarises the research findings in this area.

Table 4.5: Summary of Empirical Research on Technology Transfer

	Thursby & Kemp (2002)	Rogers et al. (2000)	Politz et al. (2000)	Siegel et al. (2003)	Thursby et al. (2001)	Carlsson & Fridh (2002)	Friedman & Silberman (2003)
Sample size (Universities)	112	131	142	113	47	170	83
Years analysed	1991-96	1996	1991-98	1991-96	1994-96	1991-95	1997-99
Measurement of effectiveness	Licenses executed; industry-sponsored research; patent applications; invention disclosures; royalties received	Scale based on invention disclosures; patent applications; licenses; license income; start-ups	Patent applications 1991-98 in biotechnology; total university patents	No of license agreements and license revenues	Licenses executed; royalties received; number of patents; amount of sponsored research	Technology transfer modelled as a sequence of events, focus on number of patents and licenses	Invention disclosures, licenses, start-us; royalty income licenses with equity
Primary data sources	AUTM	AUTM, NSF	AUTM, UPSTO	AUTM	AUTM, Survey of TTO	AUTM, Survey of TTO	AUTM

Key results	Thursby & Kemp (2002)	Rogers et al. (2000)	Politz et al. (2000)	Siegel et al. (2003)	Thursby et al. (2001)	Carlsson & Fridh (2002)	Friedman & Silberman (2003)
	Faculty quality important; private more efficient than public; medical school less efficient	Significant and positive are faculty quality; no of staff; federal research funding	Significant and positive are faculty quality; no of staff; federal research funding	Universities in States with higher levels of R&D are less inefficient; older TTOs are better	Significant and positive are invention disclosures; no of staff; medical school. Not significant is faculty quality	Research expenditures; invention disclosures; years TTO operating is important	Importance of rewards for faculty involvement; location; mission; and experience of TTO staff

Source: Based on Friedman & Silberman (2003).

In summary, technology transfer predominantly means translating research results (inventions) from the science lab to the marketplace (product) (Grady & Pratt, 2000). Success stories, mainly from the US, and the potential for alternative sources of funding as well as regional development impacts, make the technology transfer route an extremely attractive one for universities. In response, many universities have developed ILO/TTO structures to facilitate such technology transfer. Often, however, the risks in terms of the uncertainty involved in research and subsequent technology transfer are underestimated. In order that the university does not become exposed, it is essential that the best and most appropriate technology transfer mechanisms are chosen, given the expertise and resource constraints of the given university. Good institutional policies and capacity can help reduce the legal risk (for example, infringement) that is associated with technology transfer.

Mechanisms for Technology Transfer

In the following sections, we review empirical evidence and best practice activities pertaining to the main mechanisms of technology transfer used by TTOs – patenting, licensing, and company formation.

Intellectual Property

Intellectual property (IP) refers to the legal form of protection for inventions, brands, designs and creative works. The four main types of IP rights (IPRs) are patents, copyright, designs and trademarks. Most technology transfer from universities involves patents, so this chapter uses IP to refer to them. However, other forms of IP also play a role in business-university collaborations, especially in creative industries. HEIs use intellectual property protection to provide the legal fabric of property ownership that makes technology transfer through licensing possible. The key underpinning to all IP-based action should be akin that suggested by Nelson at MIT: 'I think our success, and that of other universities who have managed to stick to their core principles in the face of economic forces, comes from remembering who we are and why we are in the game'. Management of IP and technology transfer by licensing and through spinout companies has been referred to as the 'harder' side of knowledge transfer (Lundvall, 1998).

There is significant potential for transferring knowledge to business in the form of IP. Jain (2004) defines IPR to be 'policies that assign and protect the rights to earn income from innovative and creative activity These rights provide legal authority to control the dissemination and commercialisation of new information and ideas and to enforce sanctions against their unauthorized use'. He goes on to say that IPRs are controversial in nature, as they often lead to higher prices and a reduced availability of products. However, in the industrialised world, they are vital for innovative companies to survive, as they need protection from the threat of imitation. These transfers take a range of different forms and have been growing at a rapid pace in recent years. Universities are encouraged to protect their academic inventions and creative works by a general strengthening and broadening of intellectual property provision (OECD, 2003: 9). In many cases, TTOs have been developed specifically to deal with these issues.

While, in the US, legislation dictates the codes and practices in managing and transferring IP, in Ireland, and the rest of Europe, this is largely not the case. Indeed, there is agreement among research funders, researchers and industry that uncertainty about IP ownership is one of the main barriers to effective technology transfer and research collaboration. In an effort to address this issue and to provide some much needed clarity, in April 2004, ICSTI published a *National Code of Practice for Managing Intellectual Property from Publicly-funded Research* (ICSTI, 2004). The code was widely endorsed by stakeholders in industry and all the major funding organisations, the Irish Business & Employers Confederation (IBEC), and the Irish HEIs. The code incorporates issues such as:

♦ Ownership of the IP.

♦ Principles for the management of intellectual property.

♦ Code of practice for the management of intellectual property from publicly-funded research.

♦ The duty to report discoveries.

♦ The elimination or management of conflicts.

♦ Guidelines for the implementation of the code of practice.

The *National Code of Practice* only addresses IP generated from 100% publicly-funded research. IP generated from collaborative, public–private funding requires equal attention. At the request of the Tánaiste, ICSTI is currently developing a code of practice for the management and commercialisation of intellectual property from collaborative research (Forfás, 2004; ICTSI, 2004).

Patents

A patent is a legal protection granted to the owner of an invention or process. To be eligible for a patent, an invention or process must incorporate technical or functional novelty (Forfás, 2004a: 7). The level of patent registration gives one measure of the scale of knowledge creation and commercialisation. The highest levels are found in the most innovative countries such as the US, Sweden and Finland (OECD, 2003). **Table 4.6** compares the number of Irish patent applications and patents granted through the European Patent Office (EPO) compared with Finnish figures.

Table 4.6: European Patent Office Applications and Grants:
Ireland & Finland

	1999		2000		2001	
	#	%	#	%	#	%
IRELAND						
Applications	162	0.18	212	0.21	257	0.23
Granted	45	0.13	39	0.14	53	0.15
FINLAND						
Application	1,017	1.14	1,223	1.21	1,571	1.43
Granted	352	1.00	264	0.96	340	0.98

Source: Forfás (2005).

The number of patents issued to business and universities has increased rapidly in the US, EU and Japan since the mid-1980s (Hernes & Martin, 2000: 34). If HEIs are producing useful knowledge, and are to be encouraged to commercialise this knowledge, they need to have incentives to do so and the ability to protect their intellectual property. Patents are used as a means to signify university expertise, therefore patents are crucial in encouraging fruitful relationships with industry. It is crucial to acknowledge, however, that *whatever* the content of legislation or policy on intellectual property rights, a patent does not inevitably mean that an invention will be transformed into an innovation (Hernes & Martin, 2000: 34). Not all patents will be successfully commercialised, while others can have a very significant impact (Forfás, 2004a: 20). Patents are likely to be limited in their ability to signify causal links (Tijssen, 2001: 39). Nevertheless, the patterns of publications cited in patents have been used by authors as a proxy indicator of university-industry links. The extent of the growth in patenting activities by universities is evidenced by the fact that, in the US, for example, in absolute numbers, the magnitude of patenting by the universities has increased more quickly than the overall trend of general patenting (Graff *et al.*, 2001) (see **Figure 4.2**).

Figure 4.2: Patents Assigned to US Universities, 1969-1999

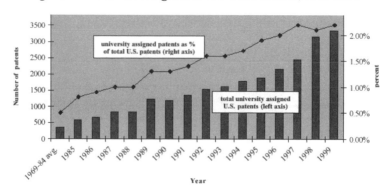

Source: US Patent & Trademark Office.

According to Henderson *et al.* (1998) the number of patents assigned to HEIs per research dollar spent at HEIs has more than tripled. Critically, however, the role of patents is not as straightforward as in companies. Jensen & Thursby (2001) report that, in just 28% of the cases, a patent issued at the same time as a license. In all other cases (72%), a patent application has been filed, but it is clear that the inventive work coming out of universities is transferred out at the pre-patenting stage. Therefore, as Graff *et al.* (2001: 19) note 'the number of granted patents is just the tip of the iceberg when it comes to indicating the value generated by university research'.

Furthermore, the number of patents filed will be a function, not only of the extent and quality of invention disclosures, but also of the resources available to universities to patent inventions. In many cases, the cost of legislative expertise and the patenting process generally reduce the likelihood of such occurrences. In an Irish context, it is estimated that there are 37 individuals registered to practice as patent agents and that the average cost of obtaining a patent for an invention is circa €100,000 (Forfás, 2004a). This is why some inventors attempt to license their technology before the major funding needs arise in the patent process. The licensor would then take over patent filing and defence costs (Forfás, 2004a: 14). In an Irish context, Enterprise Ireland has recently revised its scheme for the funding of intellectual property arising from academic research (the Patent Fund).

As of 1 January 2006, support is available for patent application costs incurred by HEIs at any (or all) stages in the process of patent protection, as follows:

♦ Stage 1: Funding of up to €7,000 to assist with the costs of preliminary patent protection.

♦ Stage 2: Funding of up to €20,000 to support patenting costs arising in the continuing prosecution of an already-filed initial patent application or the extension of patent coverage to other countries.

♦ Stage 3: Funding to provide support for the later stages of the patenting process, which can rise to €50,000.

The scheme applies irrespective of whether the research in question was originally funded by Enterprise Ireland. Such schemes provide useful mechanisms to facilitate HEIs in dealing with the expense of the patenting process and, by so doing, should encourage technology transfer and further commercialisation of research.

A weakness noted by Allan (2001) is that some institutions invest heavily in building up a portfolio of patents to the neglect of marketing and licensing them. In some cases, this has been attributed to marketing and commerce skill deficiencies among the staff of TTOs.

Invention Disclosure & Deciding What to Patent

The patent system provides a means for inventors to protect the fruits of their labours, and thereby attract more funding, talented research assistants, and notoriety (Monaq.com). Patents are also seen as a common indicator of sensitivity to industry needs (Plonski, 2000). The extent of patenting activities, however, is a function of the amount of invention disclosures (the raw material for technology transfer) submitted to the TTO (Siegel *et al.*, 2003). In the US, in 2002, AUTM data reveals that 15,573 invention disclosures were received by 221 institutions, an increase of 14.8% from the previous year (see **Figure 4.3**).

TTO Key Challenges

The key challenge for TTOs is separating out commercially-promising ideas from the large volume of publicly-orientated knowledge constantly being created at the HEI (Scott *et al.*, 2001: 16). Graff *et al.*, (2001) note that the probability of patenting is greater if certain conditions are met (see **Table 4.7**).

If none of these conditions are met, the TTO may chose not to patent the invention but rather give the faculty inventor the right to use and pursue it. Some faculty have used unpatented technology to start firms. While a short delay in publication may be imposed for the practical purpose of allowing a priority date to a patent application, faculty members are permitted and encouraged to publish the results and to compete in the academic research race (Postlewait *et al.*, 1993).

Figure 4.3: Invention Disclosures Received by US Universities, 2002

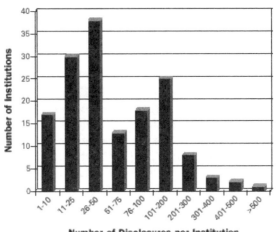

Number of Disclosures per Institution

Source: AUTM (2002).

Table 4.7: Conditions Influencing the Probability that TTO will Patent

The probability of patenting is greater when:

♦ A technology is clearly patentable.

♦ A technology has favourable cost-benefit considerations (good commercial prospects or potential buyer already interested in it).

♦ A potential buyer offers to patent collaboratively.

♦ The OTT is approached, and even persuaded, by a faculty inventor.

Allan's review of TTO best practice reveals that there is little correlation between budgets, number of patents, or other quantitative measures and the degree of success a TTO will have in terms of generated income. Columbia University, for example, has often appeared at or near the top of AUTM's survey in terms of licensing income generated. Yet, it was granted an average of only 34 patents per year from 1994 to 1998.

Yale University reviewed its 850 invention disclosures from 1982 through 1996 and learned the following:

♦ 1% (10 of 850) of total disclosures led to 70% of $20.4 million received.

♦ 4% (33 of 850) of disclosures accounted for 90% of the total licensing income.

♦ 88% (748 of 850) of disclosures generated less than $10,000 each, the approximate cost for processing one invention disclosure.

Allocation of Resources

These examples suggest that a key focus for TTOs is to decide how to allocate development efforts, since not all disclosures offer equal promise of success (Allan, 2001). Successful patenting and subsequent commercialisation generally will also require the active involvement of HEI inventors (Goldfarb & Henrekson, 2002). In the US, ideas reach TTOs in primitive states and much critical knowledge is often tacit. Thus, it is clear that establishing the value of an innovative idea is a complex operation. Even the brightest idea, without appropriate technical development and marketing, has little or no chance of success (Fassin, 2000: 35). In order to address these issues, increasing funding has been allocated to the proof of concept stage. This is also an important criterion to have in place to ensure funding from government agencies.

Table 4.8: Proof of Concept Fund: Scotland

The Proof of Concept Fund was launched as a three-year Stg £11m pilot initiative in October 1999 by the Scottish Executive and is administered by Scottish Enterprise, the Government-funded economic development agency for north east, central and southern Scotland. The fund was a recommendation of the Knowledge Economy Taskforce Report. The purpose of the fund is to address the development gap between scientific discovery and the proof of concept or prototype stage at which normal commercial types of funding can be accessed. It is allocated through competitions to higher education institutions and research institutes for projects at the pre-development conceptual stage.

The fund is directed towards a proof of concept activity that supports the development of clusters of companies and organisations in related industries which have economic links including skills and infra-structural needs. The clusters are biotechnology, semiconductors/microelectronics, optoelectronics, food and drink, oil and gas and creative industries (media). The funding levels are from Stg £50,000 to Stg £500,000 with one call for proposals in 2000 and two per year thereafter. Successful proposals were made in the first call in the areas of biotechnology and semiconductors/microelectronics.

Source: Lambert (2003).

The OECD survey shows that European HEIs tend to file most of their patents in their home country, while fewer academic patents are filed at European level or overseas (2003: 45). This reflects the importance of filing a patent within home jurisdictions first, but there are concerns that subsequent patenting at the EU level could be deterred by the cost of an EPC patent.

Table 4.9 Proof of Concept: Enterprise Ireland

Enterprise Ireland's Commercialisation Fund (proof of concept) was launched in 2003. The programme aims to 'support academic researchers in establishing that a scientific concept, from whatever source, is sufficiently robust, is seen to address a viable market and is not encumbered by intellectual property considerations'. The scheme focuses on a proof of concept model, where individuals or small groups work on short applied projects to develop a product concept through to a stage where a route to commercialisation is clear, either as a campus company or through licensing.

Under the programme, grants are awarded to researchers in the third-level institution to develop and examine an idea/concept to establish the scientific/technical merit and feasibility of the work proposed. This fund can be used by a researcher to determine a clear route to commercialisation, including the creation of new high potential start-up companies or technology licensing agreements. Enterprise Ireland pays 100% of the total eligible costs within third-level institutions, with projects falling in the region of €50,000 to €90,000. Projects are usually undertaken within a 12-18 month period.

Over the past few years, the number of projects supported annually was around 60. Panels of experts covering a range of disciplines, and drawn from the academic and business worlds, assess proposals.

Source: Arnold et al. *(2004).*

LICENSING: ACTIVITIES & PRACTICES

When university research leads to technical innovations, one route to commercialisation is to license the technology to other firms, rather than try to exploit the technology in a direct way. Licensing agreements typically involve selling to a company the rights to use a university's inventions in return for revenue in the form of up-front fees at the time of closing the deal, and annual, ongoing royalty payments that are contingent upon the commercial success of the technology in the marketplace. License fees typically range from $10,000 to $50,000 but may be as high as $250,000, while royalty rates are typically 2% to 5%, but may be as high as 15% (Feldman *et al.,* 2002).

Siegel *et al.* (2003: 31) found that, despite the perception that there are multiple outputs of technology transfer, in reality, licensing activity is 'by far the most crucial output'. The key questions in licensing are: 'which areas of university research are most likely to lead to licensing potential; how can universities best organise their licensing activities; and how much revenue from licenses can be expected' (Scott *et al.*, 2001). It is not necessarily the stronger patents that are the object of licenses; early stage-technologies, know-how and materials are equally licensable (OECD, 2003: 67).

In absolute terms, US universities generated the largest amount of income from licenses, over \$1.2 billion, followed by Germany at €6.6 billion. Per institution, this can range from thousands to millions. Data on licensing revenue per license reveals the skewed nature of income from technology transfer.

The Value of Licensing

The utility of considering licensing is that the licensing of intellectual property creates commercial value: licensing information is likely to be a better indicator of technology transfer activity than intellectual property rights alone. Also, crucially, the terms of a licensing contract can be construed so as to achieve non-commercial goals. According to the OECD (2003), 'clauses can be included to ensure that technologies are broadly disseminated or exploited domestically. Alternatively, it can be stipulated that PROs can continue using the IP, thus safeguarding against fears that commercial ties will interfere with advances in research'.

While HEIs can receive more than three times as much revenue from corporate-sponsored research as they do from licensing income, the key advantage to licensing is that income thus generated is uncommitted. This means that, rather than having the direction of spending imposed, as is the case for corporate-sponsored research, HEIs have liberty to use the income from licensing to support its research and education in any way it chooses. Licensing thus can serve to reduce criticisms of the negative effective of commercialisation activities. **Figure 4.4** below shows that, in the UK, while licensing is increasingly popular, institutions are still limited in the number of license agreements they negotiate.

Licenses can be granted for the use of patented technologies, for technologies with a patent pending, for unpatented technologies (for example, biological materials or know-how) for which no formal form of protection has been or will be sought, for innovations covered by a *sui generis* form of protection (for example, plant varieties) or for creative works covered by copyright (OECD, 2003: 60).

Echoing data from the UK, country statistics drawn from the OECD survey show that two-thirds of countries report that their HEIs negotiate less then 10 licenses a year (**Figure 4.10**). The other one-third report that they negotiate between 14.7 and 45.8 licenses a year.

OECD findings indicate that technologies for which patents are pending, and non-patented innovations, are more frequently the object of a license as opposed to patented inventions. Thus, it is common for universities to license early-stage technologies to firms, which subsequently invest in their further development.

Figure 4.4: UK: Licenses & Options Executed, 2002

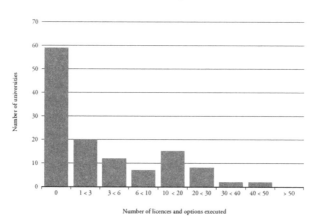

Source: UK University Commercialisation Survey Financial Year 2002 (UNICO, NUBS, AURIL, 2003).

Table 4.10: OECD: Licences *Per Annum*

Country	Total # Licenses	Average per University	# Responding Universities
Australia	234	7.1	33
Italy	27	1.4	20
Korea	44	8.8	5
Netherlands	250	14.7	17
Switzerland	200	9.5	21
United States	4,049	24.1	168

Source Adopted from OECD (2003: 61).

Table 4.11: Technology Licensing: USA

Of the 3,600 universities in the United States, some 500 are research universities. Of the total, one in 12 has a technology licensing office (TLO), giving rise to 4,800 patent applications (1998) and 350 spin-off companies per year. Revenues from technology licensing are about 2% of university revenues.

In the area of technology licensing universities have:

♦ Transparent and extensive policies.

♦ Extensive and effective contractual procedures.

♦ TLOs varying in size and competence and having professional staff.

It is a requirement for researchers to disclose inventions that may be commercialised through the office for technology licensing. An example, on a scale relevant in the Irish context, is the Georgia Institute of Technology. Typically, the researchers receive 30% to 50% of royalties. The TLO receives 15% of the gross income and, in addition, takes equity in new spin-offs based on university technology.

Licensing at MIT & Oxford

The case of MIT was explored in **Chapter 3**. MIT is a prime example of a university that has focused on licensing existing technology. It discloses about 450 inventions per year. MIT's TLO then licenses these inventions as non-exclusive or exclusive licences to industry and local venture capital firms. Similarly, Oxford University's ISIS is one of the UK's most prolific technology transfer offices when it comes to agreeing licensing deals. Since 1997, it has entered into 160 such agreements on university technology. Although it does not prescribe the balance between spin-outs and licensing in any way, over two-thirds of its technologies are licensed to existing firms. In 2003, ISIS agreed 37 licences and formed seven spin-outs. Drawing from the example of ISIS at Oxford, **Table 4.12** shows some of the benefits of licensing to industry, which include fast access to new technologies, often on an exclusive basis, leading to new products and services, increased revenue and possible employment growth.

Table 4.12: Licensing & Exploitation Companies: ISIS at Oxford

A complementary development is the creation of an exploitation company, of which ISIS in Oxford is a good example. ISIS is charged to undertake, on the university's behalf, the commercial exploitation of its research. Some universities require that all exploitation pass through such a company; others set it up in such a way that it must sell its services to the academic community, in competition with existing private sector bodies. Such a company, which will have access to investment funds contributed to, in whole or in part, by the university, will use its resources to safeguard and exploit the university's intellectual property interests in terms of patents, licenses and royalties. It will serve as a venture capital arm to invest in companies growing out of the university laboratories.

The problem for exploitation companies is the size of business they can generate. ISIS, at the University of Oxford, has world class science and technology, involving several thousand members of staff and research students, who are capable of producing a significant range of intellectual products. However, the record of other exploitation companies has not necessarily been so good, especially in universities where the science base is much smaller, or less geared to leading-edge fields such as biotechnology.

The strength of the exploitation company approach is undoubtedly the commercial attitude, in that it influences decisions on whether to invest heavily in the intellectual property area. Engaging expert advice, deciding on the defence of patents in the courts, or whether to patent ideas outside the UK, are costly matters, best decided by company executives, who are answerable to a board, rather than by university administrators, who may not be specialists in the area. Equally, companies wanting to do business with a university may consider an exploitation company to be an easier working partner than a university research office.

Source: Shattock (2001).

Licensing Revenues: Cash & Equity

Research by Jensen & Thursby (2001) showed that, of the 62 institutions they surveyed, 23% of the license agreements included equity. Yet the magnitude of HEI ownership of private equity should not be overstated, as only 3% of the total TTO revenues were in the form of equity. Findings from Castillo's *et al.* (2000) survey, shown in Table **4.13**, indicate preferences among TTOs for compensation in cash or equity of the licensing firm. While many HEIs are open to accepting equity as a form of compensation, this is largely skewed towards private HEIs. Most public HEIs do not see equity as a feasible option for payment, as this can clash with mission and objectives that emphasise the role of technology transfer in increasing 'public good' rather than maximising returns.

Table 4.13: Equity Preferences Reported by TTO Officers, 1999

Question	Public Universities %	Private Universities %
'My institution will not, or is not allowed to, consider equity'	10.0	13.3
'My institution prefers cash payments but will consider equity'	63.3	46.7
'My institution believes cash and equity are equally viable forms of payment'	26/7	40.0
'My institutions prefers equity'	0.0	0.0
Total	**100**	**100**

Source: Castillo et al. (2000).

Exclusivity of Licenses on Patent Technologies

Licenses are permissions granted by the owner of a piece of intellectual property to another party for the use of the invention or work. Licenses can be granted on an exclusive basis, to a single licensee, thus guaranteeing a strong degree of market exclusivity. Licenses can also be granted non-exclusively, to many parties, as is frequently the case for software, or limited by other means – for example, for a limited time period, in a particular geographical territory or market or technological field (OECD, 2003: 62). Companies tend to try and negotiate exclusive licenses when the future depends on the promise of a limited number of technologies.

Notably, the OECD (2003: 62) survey found that 'the extent to which universities grant exclusive, rather than non-exclusive, licenses varies widely from country to country', so that 'no best practice appears to have emerged about how to construct licensing agreements or how to balance the public good of broad diffusion of innovations with private desires for exclusive rights' (OECD, 2003: 64).

Negotiating License Agreements

When negotiating license agreements universities generally try to rigidly adhere to three guiding principles that are included as clauses in most agreements. These clauses are to:

♦ Encourage the use of publicly-funded technologies.

♦ Ensure the broad diffusion of the research results.

♦ Maximise the future IP revenue streams.

Generally, when conducting agreements, HEIs need to be satisfied that the respective company will act in good faith in its efforts to commercialise the research. Often clauses to this effect are included in contracts negotiated by universities. These ensure that the TTO's mission of exploiting research for the public good is fulfilled. Related to this, many HEIs find that clauses in respect of publication delays run counter to their public research missions. The OECD notes that, as a consequence of this, some institutions make it a policy to avoid signing publication delay clauses, or that they limit permissible delays to three to six months. Licensing strategies negotiated by HEIs therefore attempt, through clauses, to protect public interest. Licenses also often include some form of limited exclusivity (for example, by territory or field), so that technology maybe used by more than one firm (OECD, 2003: 17).

Furthermore, a key aspect of agreements that are reached through clauses is that the university can make claims for royalties on future products developed with the aid of a licensed technology. This is especially common in the life sciences or in the licensing of research tools. Research institutions may also seek right of first refusal on licenses to any patented products developed by a licensee. In summary, the usual type of strategic clauses included in licensing agreements involve 'working the invention', reach-through clauses, right to delay publication, and right of first refusal.

Licensing Income

The potential for licensing and income generation varies widely by technological fields. **Table 4.14** indicates the average number of patents by academic discipline.

Table 4.14: Licensing Revenues & Patents by Academic Field, 1999

Academic Field	Average revenues, as % of total			Average number of patents % of total		
	Public	Private	Both	Public	Private	Both
Agriculture	10.3	6.9	9.1	6.5	5.5	8.1
Engineering & Physics	19.8	32.3	24.1	32.2	38.7	34.1
Medicine	55.2	55.2	55.2	44.4	51.0	46.4
Computer Science	5.5	5.2	5.1	3.9	4.2	4.0
Other (including Chemistry)	10.3	0.5	6.6	11.2	0.5	7.4

Source: Castillo et al. *(2000).*

Biotechnology is the largest field, making up a significant part of the field of medicine, which dominates in terms of income (earning 55% of the total) while accounting for 46% all patents. An in-depth analysis of three universities, the University of California, Stanford and Columbia found similar patterns (Mowery *et al.*, 2001). Biotechnology tends to be early-phase technology as opposed to engineering and computers, which are subject to more mature competitive environments characterised by 'process developments' (Graff *et al.*, 2001). HEIs are recognising these trends and, through strategic technology platforms, are directing research to areas that have a greater potential for commercialisation success. HEI royalties should be higher, particularly in situations where the university has provided a more commercially-developed product and thus has contributed more to reaching the market (Graff *et al.*, 2001: 21). **Table 4.15** shows royalty rates in a variety of technology fields.

Table 4.15: Key Indicators of Earnings by Field of Technology, 1999

Type of Product	Average royalty as % of sales	Average value of up-front fixed fee ($)	Average value minimum annual royalty payment ($)
Agricultural	3.9	20,105	6,928
Engineering	6.3	32,236	16,397
Medical (therapeutics)	6.3	98,437	83,010
Medical (Diagnostics)	6.6	36,906	46,227
Medical (Devices)	6.6	37,115	38,775
Medical (Material and Reagents)	9.4	12,942	4,444
Other (includes chemicals)	7.63	78,583	42,687
All Fields	6.6	45,189	34,066

Source: Castillo et al. *(2000).*

The high royalty rates for medical research reflect the relatively high contribution of the HEI research to the value of marketed products. While therapeutic license contracts have high up-front fixed fees, there does not seem to be a correlation between the up-front fee and the annual royalty payment. This suggests that there is independence between technology fields. Notably, even for the top 10 technology licensing HEIs in 1995, the total technology transfer earnings average just 2.5% of the university research budgets. Midrange institutions generated revenues averaging just 1.5% of their research budgets, and the bottom 10% generated less than 1% (Jensen & Thursby, 2001). Note that these figures are drawn from the US, which is more advanced in terms of technology transfer; one would expect figures in Europe to be well below those cited for the US. The general point is that 'it is clear that technology transfer revenues do not pay for HEI research. Furthermore, much of this money does not return directly to the HEI's research programs but instead goes towards administrative costs' (Castillo *et al.*, 2000).

Licensing: A Cautionary Note

Only a few universities to date manage to earn a significant percentage of their total research expenditures in licensing revenues (Dartmouth, Columbia, Florida State and Brigham Young). The vast majority of US universities, including Stanford and MIT, earn less than 10% of their research expenditures from IP commercialisation. Excluding the most successful universities, the reality is that licensing income is an extra benefit, not a replacement for the major public and private sources of IP funding (OECD, 2003: 71). OECD findings show that somewhere between 20% and 40% of patents are licensed and only about half of these licenses, or around 10% of the patent portfolio, earn income (OECD, 2003: 72).

COMPANY FORMATION: START-UPS & SPIN-OFFS

Start-ups and spin-offs are the most cited form of technology transfer, as they are direct manifestation of such activity and provide a basis for measuring the direct impact of technology transfer activities (for example, numbers employed, revenues, and value of equity). Approximately 12% of university-assigned inventions are transferred to the private sector through the founding of new organizations (AUTM, 2002). Much policy attention has been drawn to the role of spin-offs and start-ups, given their potential to generate employment and stimulate economic growth. Indeed, the Fifth Framework programme in the EU envisions a significant role for universities in encouraging the creation of high-tech start-ups (Gulbrandsen & Etzkowitz, 1999).

Although the terms are often used interchangeably, it is useful to distinguish between the two mechanisms of company formation available to universities. A 'spin-off' (or 'spin-out') is a company that includes among its founding members a person affiliated with the university, while a 'start-up' firm is one that is not founded by a staff member of the university but is developing technology originating at the university (licensed technology) (Graff *et al.*, 2001). The fact that corporations tend not to give university innovations a very enthusiastic reception has encouraged universities to seek such alternative avenues for the commercialisation of promising university inventions (Scott *et al.*, 2001).

Figure 4.5: Start-up Companies Formed by US Universities, 2002

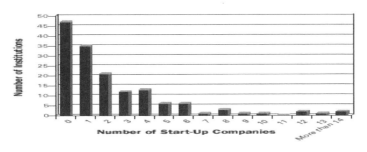

Source: AUTM (2002).

Empirical Evidence on University Start-up & Spin-off Activity

Data from the AUTM survey (2002) indicates that 450 new companies based on academic discovery were formed in the fiscal year 2002 (down 8.9% from the formations reported in fiscal year 2001). Some 83% of new companies were located in the state of the academic institution where the technology was created. Since 1980, a total of 4,320 new companies have been formed, based on a license from an academic institution. Most of these (2,741; 63%) were still operating as of the end of the fiscal year 2002. On average, academic institutions received an equity interest in 70% of their start-ups. The OECD TTO survey (2003) found similar findings. In half the reporting countries, the average number of new spin-offs or start-ups reported per HEI is under one a year. This figure is in contrast to the US, where the average is two start-ups in 2000, a figure that has remained stable since the end of the 1990s (AUTM).

Table 4.16: Spin-offs & Start Ups per TTO Reported in the Last Year

COUNTRY	Spin-offs		Start-ups		Total	Total
	Total	Per TTO	Total	Per TTO	Total	Average per TTO
Australia	-	-	-	-	32	-
Belgium*[15]	11	1.83	4	0.67	15	1.25
Germany	28	1.12	9	0.6	37	.74
Italy	14	.5	13	.46	27	.48
Japan*	1	.1	5	.45	6	.28
Korea	10	2.5	3	2.25	19	2.38
Netherlands	23	1.77	4	.36	27	1.07
Norway	15	5	1	.5	16	2.75
Russia*	8	1.33	7	1.75	15	1.54
Spain*	8	.67	3	.3	11	.48
Switzerland	39	1.77	5	1	12	2
US	-	-	390	2	-	2

* University & PRO.

Source: Adopted from OECD (2003).

This empirical data serves to highlight the relatively few companies that are formed as a result of a HEI invention. The attention such activities attract from the media often suggests otherwise.

Stimulating Spin-off Enterprises: University of Louvain La Neuve

The University of Louvain La Neuve in Belgium has a long-established science park and a good track record in fostering industry-academic links. In 1994, it was centrally involved in an EU project, 'Transfert', which brought the university together with intermediaries involved with developing linkages with industry.

---

[15]   Note Belgium = Flanders only.

Table 4.17: University of Louvain La Neuve

The R&D Liaison Unit of the University and Promotech, a Business & Innovation Centre (BIC) located in the Nancy region of France, developed a project that used intermediaries, such as BICs, to help the commercialisation of university research. One particular area is in providing a link between potential university spin-offs and venture capitalists. At this stage, speed is considered important, in case the spin-off locates outside the region or fails to get off the ground as a result of delays putting financial resources together. Other areas of operation are in the field of effective project identification, determining the most effective exploitation and commercialisation route for research, and in improving patenting and licensing services. The project has also helped to develop a wider and more supportive regional framework for university research commercialisation, which has allowed the university to interact better with a range of local business and technology agencies and intermediaries.

The frequency of TTO start-up and spin-off activity varies significantly across countries and between universities. The OECD notes that Korea, for example, has historically placed too much emphasis on such activities. The Dutch example is insightful, as it indicates the importance of sound management of technology transfer geared towards exploiting IP through start-ups. The Dutch experience highlights the importance of a mature programme, which possesses not only sufficient IP and legal know-how, but also draws insights from other disciplines in order to develop an entrepreneurial concern. In Ireland, an effort to promote company formation in the form of a technology transfer centre, NovaUCD, was recently unveiled. This is a the purpose-built centre, which offers a supportive environment and incubation facility to assist innovators and entrepreneurs in taking their ideas from proof of principle to full commercial success (Cope, 2004: 13).

Some HEIs routinely transfer their technology through the formation of new firms, while other HEIs, such as Columbia University, rarely generate start-ups. Moreover, rates of start-up activity are not a simple function of the magnitude of sponsored research funding or the quantity of inventions created. For example, Stanford University, with sponsored research expenditures of $391 million, generated 25 TLO start-ups in 1997, whereas Duke University, with sponsored research expenditures of US$361 million, generated none. In the following section, we illustrate some studies that have attempted to explain the cross-university variation in TLO company formation activities.

Factors Influencing University Company Formation

A study by Di Gregorio & Shane (2003) compared four different explanations for cross-institutional variation in new firm formation rates from HEI TTOs. These variables included the availability of venture capital in the university area; the commercial orientation of university research and development; intellectual eminence of the university; and university policies. The methodology employed involved using negative binomial models in generalised estimating equations (GEE), which are an extension of generalised linear models applied to longitudinal data. Control variables included number of inventions, number of TLO staff, and sponsored research expenditures. The sample involved looking at company formation activities of 101 universities between 1994 and 1998. The research found that important influences with respect to company formation activities included:

♦ **Intellectual eminence:** The HEI's prestige or reputation makes it easier for researchers from more eminent universities to start companies to exploit their inventions than researchers from less eminent universities. It was found that:

  ◊ Better quality researchers are more likely to start firms to exploit their inventions than lesser quality researchers; and on average, higher quality researchers are found in more eminent universities.

  ◊ Industry tends to fund less risky research, and the perceived quality of research is directly associated with the reputation and intellectual eminence of the university.

♦ **TLO policy on equity:** The study found that certain policies could generate more TLO formation activity, because those policies provide greater incentives for entrepreneurial activity. The HEI's willingness to take an equity stake in TLO company formations in exchange for paying patenting, marketing, or other up-front costs facilitated the formation of companies. *Ceteris paribus*, it was found that HEIs that have previously demonstrated a willingness to take an equity stake in licensees in exchange for paying up-front patenting and licensing expenses have a start-up rate that is 1.89 times that of universities that have not demonstrated a willingness to take equity.

♦ **Royalties:** The inventor's share of royalties, in the distribution of royalties between inventors and the university, influenced the propensity of entrepreneurs to found firms to exploit university inventions.

The findings also indicated that the level of commercially-oriented research had little effect on formation activities. Interestingly, the availability of local venture capital funding which has often been cited in other studies, was not seen to have an effect on formation activities. Shane (2003) examined the history of each of MIT's patents and discovered that 26% were fully licensed to a new firm specifically set up to exploit the technology. This reflects the MIT TTO's pro-start-up policy.

Shane (2003) found that five key dimensions determine whether a new invention will be commercialised by a start-up:

♦ Observability and tacitness of knowledge in use.

♦ The age of the field.

♦ The tendency of the market towards segmentation.

♦ The effectiveness of patents.

Institutional Equity Holdings

Unlike established firms, new firms lack cash-flow from existing operations, making them cash-constrained. University equity investments made in lieu of paying patent costs or up-front license fees to universities reduce the cash expenditures of new firms, thereby facilitating firm formation (Gregorio & Shane, 2003). The tendency to consider equity options in new start-ups stems from concerns that traditional licensing as a technology transfer mechanisms has not yielded major financial returns for most institutions (Feldman, 2002). Further, pursuit of traditional methods mean that lag times may exist before the receipt of any royalties by the university. Equity, in contrast, may provide a financial return in the case of an Initial Public Offering (IPO) or an acquisition by another firm. University equity investments also mean that universities share in the fortunes of a firm rather than in just the fortunes of a technology, of particular benefit when subsequent products may embody knowledge first diffused as part of the initial transfer. According to the AUTM, in the fiscal year 2002, 203 institutions reported that they had granted 313 licenses with equity to start-up companies, down 10% from 2001. The total number of licenses/options with equity was substantially higher, 443 granted by 218 institutions. In 2002, 70% of the licenses with equity were licenses to start-up companies.

The subject of institutional equity holdings, however, is one that draws much controversy. Obviously, the HEI is showing faith by investing and is also seeking to realise the intellectual property rights it can legitimately claim if the research has been developed on its own premises by its own staff. On the other hand, however, start-up companies normally require successive tranches of investment to turn a research finding into a profitable business and skills in management and marketing become more important than the value of the original research. As a result, the HEI can often find itself being sucked into successive rounds of investment without the secure prospect of any financial returns (Scott *et al.*, 2001; Lambert, 2003). Further, investment will always be subject to market forces, and the recent sharp decline in high-tech stock prices and in IPO share prices clearly reduce the allure of the equity options (Feldman, 2002). Experience from Warwick University is instructive.

Table 4.18: Warwick University: Institutional Equity Policy

Warwick made an investment in a biotechnology company launched some 15 years ago. The company is still alive and has secured successive rounds of venture capital funding, so that its value is much higher than it was initially, but the university's equity share has fallen to a very small proportion of the total share value. If the company were to be sold, the university might realise its investment, but it is unlikely. And even if the company goes into production and becomes a really profitable business (which the initial research idea suggested it certainly would do), the odds are against financial return to the university being better than if it had invested the sum in more conventional ways.

Source: Adapted from Lambert (2003).

Best Practice for Universities in Managing the Formation of Spin-Offs

Howells & MacKinlay's (1999) survey of academics in 10 universities in Europe identified the following basic elements of good practice for universities in managing the formation of spin-off companies:

♦ **Provide clear and transparent guidelines** to all HEI staff wishing to set up a spin-off company, so that the prospective entrepreneurs have a solid framework before they make their judgement about whether to proceed with setting up the company and how this should be done.

♦ **Proactively provide information**, contacts and support on how to establish a company.

♦ If the HEI owns the IP, which protects the innovation or technology that forms the basis for the company, it needs to **jointly discuss and decide at the outset** with the academic inventors or prospective entrepreneurs whether setting up a company is the best option.

♦ Prospective entrepreneurs should be allowed **to remain on a part-time research or teaching position with the HEI** if they so desire, although there should be clear demarcation of responsibilities and practices between the two jobs.

♦ **Provision of advice to prospective entrepreneurs** on what is required to effectively operate and run the company and what additional manpower expertise may be needed to be recruited to fill the gaps.

♦ Help **develop business plans to define clear commercial goals** for the new company as well as addressing market research and sales strategies.

Supporting Spin-off Enterprises: University of Twente

As part of an initiative to help new university-based spin-offs, the University of Twente in the Netherlands created the Temporary Entrepreneurial Post scheme. This scheme is centred on graduate engineers and seeks to provide a bridge or support scheme for the key transition period when a new graduate is setting up a company and testing the market for their idea. To do this, the university offers a part-time research assistant post to the potential entrepreneur to provide the person with a basic income, as well as to allow some spare time for the person to set up their enterprise. The potential entrepreneur is also eligible for grants to help set up the company itself, and advice and guidance is also provided through the university's industrial liaison office, together with information and introductions to venture capital firms. The university offers 20 such posts *per annum* and the drop out rate is less than 25%. It is estimated that some 300 spin-off companies have been created since the scheme was established in 1979.

Table 4.19: Twente University: The Support Package

The package includes:

♦ A part-time post in one of the departments at the university.

♦ Expert advice and support by colleagues in the department.

♦ The use of the university's facilities, such as laboratories and test equipment.

♦ Basic accommodation and office facilities within the university.

♦ A risk-bearing loan with no interest.

♦ Use of the university's image and contacts within its wider network.

♦ Possibilities of acquiring orders through the industrial liaison office.

♦ Business management support and practical guidance by an experienced entrepreneur or mentor.

♦ Support in the further development of a business plan via the school course in entrepreneurship.

University Company Formation: A Cautionary Note

While company formation is the most direct manifestation of technology transfer, TTOs must exercise extreme caution when engaging in such activities. As has been illustrated, success in this area stems from sound management, vast experience and not only sufficient IP and legal know-how but also insights from multiple disciplines. These difficulties need to be appreciated by TTOs. For example, some TTOs purposefully build up expertise in this area and follow an explicit company formation strategy that leverages this knowledge. Other TTOs recognise the complexities involved and only on rare occasions support company formation (for example, Columbia, John Hopkins).

Also, while one or two cases are constantly referred to, the returns from company formation activities have not been dramatic. Often such formation activities place strains on already stretched resources. This concern was voiced by one of our interviewees: 'Universities shouldn't be managing companies, that's not what we're good at, we're good at research and we should be commercialising those opportunities externally, not trying to build companies internally. Even though one or two universities have developed adequate management systems to do that, but that's a big cost as well in terms of infrastructure and you'd have to wonder whether that's worthwhile, it certainly isn't in the case of smaller university'.

A widely-held view in business and universities is that too many spin-outs have been created in the last number of years and that a large number of them will not succeed in the long-term. The quality of spin-outs varies widely across different universities, and is contingent on the supporting infrastructure as discussed above. The Lambert Report (2003) noted that the best way to judge quality is by looking at the ability of a spin-out to attract external private equity. This indicates whether there is real market interest in the new company. Referring to the UK, Lambert (2003) noted that at one end, Oxford University has attracted private capital to 95% of its spin-outs since 1997 but almost a third of the universities that created spin-outs in 2002 did not bring in external equity for any of their new companies. The report concluded that there has been too little licensing and too many unsustainable spinouts (see **Figure 4.6**).

Figure 4.6: Spin-outs Created & Licence Income in 2002

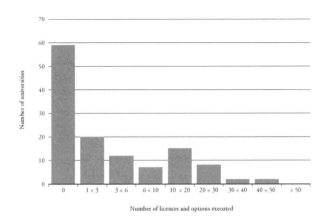

Number of licences and options executed

Source: UK University Commercialisation Survey Financial Year 2002 (UNICO, NUBS, AURIL, 2003).

OTHER TECHNOLOGY TRANSFER MECHANISMS

Non-Patent IP Actions

Applying for patents is only one of a range of actions that TTOs can take to help to protect and exploit their institution's intellectual property. In the US, some TTOs are also involved in registering copyright to protect creative works such as software, databases, educational materials (papers, books, courses) and multimedia works (on-line courses, digital text books). It is only recently that universities in some countries have begun to claim rights to copyrighted works. Two sorts of *sui generis* protection – industrial design registration and plant breeder's rights – might also be managed by the TTO.

Unlike patent inventions, ownership of copyright works in particular is often neglected. Yet universities generate a lot of their IP in the form of literary or artistic works that can be protected by copyright. Ownership of copyright for publicly-funded databases and other universities has become an important issue, as there are increasing demands by firms and the public for access to such databases. PROs in the US and the UK, but also in Ireland and France, are paying more attention to clarifying copyright ownership of works created by their employees, particularly in light of Internet publication and access.

Ownership of IP Resulting from Research Sponsored by Industry

The results of the OECD (2003) TTO survey suggests that, in most OECD countries, ownership of IP arising from research at PROs that is jointly funded by industry is shared according to the following principle: ownership of the patented invention remains with the PRO, while the firm funding the research retains the right (or option) to license the patent on an exclusive basis. When a firm funds more than 50% of the research, it may claim title to the IP. In most cases, it is possible to construct arrangements that can serve the commercialisation needs of companies, while still vesting intellectual property ownership with the university. This situation does not necessarily hold true in the case of copyrights.

When collaboration negotiations are combined with a contentious licensing negotiation, they can be much more arduous. Therefore, collaboration partners often try to resolve commercialisation terms quickly or, if that is not possible, to defer the negotiation of licensing royalty rates until the research is complete.

Agreements on Background Rights

These include provisions that the university offer a good faith or reasonable effort to find potential conflicts, although these phrases can be open to legal interpretation. For these and other reasons, universities rarely sign binding agreements on background rights. Until now, there have been few instances in which background rights have become a major problem, but the issue may have a major effect in the future.

Research Tools

Some companies have expressed concern that universities are trying to patent more intellectual property than necessary, particularly in the area of research. Research tools can be highly complex entities that themselves require research to develop, and access to publicly-funded research tools is becoming one of the most contentious areas of university-industry relationships. The issue is whether these research tools will be licensed broadly or exclusively to one company, frequently a faculty start-up. The friction generated by this conflict poses a serious risk of souring the relationship between universities and companies. Clearly, companies would like to see these tools licensed broadly.

SOFT METHODS OF TECHNOLOGY TRANSFER

IP management involves carefully managing and diffusing newly-generated knowledge into society. This is often understood simply as patenting and licensing. Most PROs transfer their knowledge in various ways, including spin-offs, taking equity in start-ups, helping out SMEs, providing training courses and distributing written information. Scott *et al.* (2001) note that the best form of knowledge transfer comes when a talented researcher moves out of the university and into business, or *vice versa*. Given its intangible nature, knowledge is best transferred by people. One programme that served to increase university-industry collaboration is the linkages between Edinburgh Crystal, Wolverhampton University and Edinburgh College of Art.

Table 4.20: Case Study: Edinburgh Crystal, Wolverhampton University & Edinburgh College of Art

Edinburgh Crystal knew that, to guarantee its long-term survival, it would have to find a younger market. That meant new designs. At the same time, Wolverhampton University's School of Art & Design was looking for ways to develop links with industry. When the School's Head of Glass met Edinburgh Crystal management at a trade event in 1994, it marked the start of a fruitful collaboration. With Edinburgh College of Art also involved, the Edinburgh Crystal Masters Design Scholarship programme was created. The students work at the company full-time on 12- to 15-month placements. The company contributes to bursary funding and academic supervision costs. A three-monthly review process culminates in an external exam leading to the Masters degree.

Edinburgh Crystal's operations director said: 'We wanted to get a more contemporary feel to our glass design in a market that was then still very traditional. We felt this programme would impact positively on our design department and stimulate the product development process. This continual stream of students encourages us to push design and production possibilities, keeping us highly competitive'. Students' work has fed directly into the company's branded range, The Edge, launched nearly three years ago, and several Masters students have gone on to become full-time employees.

The School's Head of Glass said: 'It's almost impossible for an academic programme to simulate a commercial creative environment but, in collaborations like this, students are exposed to market realities. They mature quickly as designers'.

Internships

Internships are the most traditional type of linkage. However, they often encounter numerous problems if they are not managed and well-organised. BITS in India has developed an innovative programme of so-called 'practice schools' that follow a similar scheme to those offered by MIT. Practice schools are established in a number of enterprises that agree to collaborate with BITS staff on a regular basis. At the same time, students join practice schools on pre-agreed tasks and they are supervised jointly by BITS teaching staff posted at the enterprise and staff at the enterprise.

Networks & Partnerships

Forums that bring academics and business people together are likely to increase the chance that people with common interests and goals will find innovative ways to develop partnerships. The Connect Midlands mechanisms shown in **Table 4.21** are useful in this regard.

Table 4.21: Connect Midlands

In order to gain critical mass, a novel approach to a 'companies fair' was initiated in the Midlands of England. Run by Warwick University, it involved all the major 12 universities in the area pooling their technology spin-off companies into a network that could more efficiently seek investors and expertise. 'Connect Midlands' is being used as a platform or showcase for spin-off university companies to showcase their wares for potential investors. This network is based on the Californian model and its mission is 'to nurture the development and growth of technology-related enterprise in the Midlands by connecting entrepreneurs to the resources they need to succeed'. Its main focus involves making investors aware of Midlands-based investment opportunities through the use of a rigorous screening process that redirects the unsuccessful applicants to another programme called 'InvoRed', which provides them with training, workshops and coaching so they may attain the standards necessary in the future to be included in the Connect showcase.

Those companies deemed successful enough to be included in Connect are placed into four different sections, depending on their stage of development and funding requirements: the *development board* exists to help SMEs receive appraisal from a panel of experts who can help the company focus on key profiles; the *springboard conference* consists of 12 to 15 SMEs seeking up to £0.5 million to pitch to an audience of investors and intermediaries; the *roundtable* involves 3 to 4 SMEs seeking early stage of finance to present to 12 target investors over an exclusive lunch or dinner; the *investment conference* allows 15 to 18 SMEs seeking up to £3 million to pitch to an audience of investors and intermediaries.

The main advantage of such a programme is that it creates an effective marketplace, where companies can show their wares to potential investors who have been assembled with the motive of potential investment. Although Warwick University needs this critical mass to make its spin-off companies appeal to investors (as an investor might not be interested in going to Warwick to look at just one company), it can be seen to benefit from running this network by being at the core of the decision-making process. It also receives funding for running the network.

An example of the triple helix system in action is collaborative initiatives between sectors in terms of education and research. Generally, such co-operation between parties is seen as the mechanism that drives forward successful commercialisation and technology transfer activities.

Table 4.22: The Teaching Company Scheme in the UK's Knowledge Transfer Partnerships

At the heart of each partnership is one or more KTP associates, a high-calibre graduate who is recruited to work in a business on a project that is central to its strategic development. A project may last from 12 to 36 months. The university partner provides its expertise and jointly supervises the project together with a representative from the company. The costs are part-funded by Government, with the balance being borne by the participating business. The total investment by Government in the Teaching Company Scheme was £25m in 2002-03.

An evaluation of TCS was undertaken for the Government in 2001. Based on a sample of interviews with university and business partners involved in TCS programmes, it concluded that:

♦ 44% of business partners had not previously collaborated with a university.

♦ 75% of businesses regarded the programme as strategic to their business.

♦ 38% of businesses introduced a new technology and a further 45% introduced a significant advance in technology.

♦ 50% of the companies interviewed expected their programme to have a positive effect on future sales and profitability.

♦ 54% of associates stay with the company.

♦ 94% of businesses would recommend the TCS to other companies.

Other mechanisms of softer technology transfer include academics taking places on company boards or leveraging existing alumni networks. As noted by the Lambert Report (2003), universities, departments and faculties should develop their alumni networks in order to build closer relationships with their graduates working in the business community. Other efforts include educational offerings as highlighted in **Table 4.23**.

Table 4.23: Entrepreneurship Course: USA

California Institute of Technology (CalTech) provides a course on entrepreneurship, open to both undergraduate and postgraduate students. Essentially, the course contains all the elements critical to building a business and to understanding the language of business.

The topics covered include:

- CalTech's disclosure and patent policies.
- CalTech's incentives and grants, technology transfer, and deals.
- Patents and intellectual property.
- Strategic role of patents and litigation.
- Marketing in technology-based enterprises.
- Formation of corporations and types of partnerships.
- Options and other financial instruments; vesting.
- Accounting and financial reports.
- Venture capital.
- Financial control.
- Initial public offerings.
- Human resources issues, job descriptions, and compensation.
- Benefits, pensions, and litigation risks.
- Business plans.
- Mergers and acquisitions.
- Corporate finance overview.
- Corporate partnerships and strategic alliances.
- Dealing with Wall Street.

Consultancy Activity

Consultancy takes the form of expert advice or analysis services. In practice, the difference between consultancy and contract research is blurred – but the general distinction is that, in consultancy, the academic provides advice to the business rather than actually conducting research. Universities have freedom to determine their policies and, consequently, there is a wide variation in current practice. Institutional limits on the time academics are allowed to spend on consultancy range from 20 days (Oxford Brookes) to 50 (Aston, Swansea) per year. Some of the more research-intensive universities set their limits at 30 days (Imperial, Leeds), while others do not have a precise figure at all (King's College London, Bristol) but make it clear that academic duties and the university's interests must be put first (Lambert, 2003).

Table 4.24: Consultancy Case Study: US Policies – MIT

Consultancy is much more of a core activity at MIT than it is in European universities. The opportunity to perform consulting work is built into its faculty employment contract, which only covers nine months of the year. The rest of the time can be filled by consultancy work. MIT provides strong financial incentives to academics to bring in industrial research income. It also removes teaching responsibilities for those who bring in more than $2m, and administrative responsibilities for more than $4m. MIT recognises the need for clear policies to avoid conflicts of interest within this framework.

Source: Lambert (2003).

IMPLICATIONS FOR HEIS

As a *caveat* to exploring the mechanisms for technology transfer, it is useful to appreciate that 'there are important field-specific and invention-specific differences in the technology transfer process and the role of patents and licenses in this process. There is substantial variation in the importance of patents and licenses, the role of university, the importance and involvement of the academic inventor, and even in the directionality and characteristics of the knowledge flows between university and industry' (Mowery *et al.*, 2004: 153).

Nonetheless, commonalities in ensuring effective technology transfer include:

♦ Clarity in the frameworks and procedures of exploitation and commercialisation policy.

♦ Transparency in decision-making concerning exploitation and commercialisation activity.

♦ Good information and contact provision to academic personnel regarding intellectual property and commercialisation procedures.

♦ Streamlined decision-making procedures.

To prevent HEIs becoming exposed to the risks associated with technology transfer, it is essential that the best and most appropriate technology transfer mechanisms are chosen, given the expertise and resource constraints of their context.

In dealing with patenting, the probability of success is greater when:

♦ A technology is clearly patentable.

♦ A technology has favourable cost-benefit considerations (good commercial prospects or a potential buyer is already interested in the technology).

+ A potential buyer offers to patent collaboratively.
+ The TTO is approached, and even persuaded, by a faculty inventor.

In term of licensing, the key questions for HEIs are:
+ What areas are most likely to lead to licensing potential?
+ How should they best organise their licensing activities ?
+ What is the expected revenue?

International evidence demonstrates that only a small number of HEIs have managed to earn a significant percentage of their total research expenditure in licensing revenues. The reality for most HEIs is that licensing revenues is an extra benefit, not a replacement for major public and private sources of IP funding.

Finally, in relation to company formations, providing clear and transparent guidelines to staff is essential. TTOs should proactively provide information on company formation. For HEIs, the key strategic issues include whether they should get deeply involved in company formations, whether they should manage company formation externally, given resource constraints of TTOs, or whether they should focus exclusively on licensing.

For HEIs, these issues are better considered in tandem with an appreciation of the stimulants and barriers to technology transfer specific to their situation, as discussed in the next chapter.

5

DEVELOPING TECHNOLOGY TRANSFER: STIMULANTS & BARRIERS

In general, the process of commercialising intellectual property is very complex, highly risky, takes a long time, and costs much more than you think it will.
US Congress, Committee on Science & Technology (1985: 12)

INTRODUCTION

While figures documenting the impact of commercialisation highlight the benefits of exploiting research, this is often done at the expense of acknowledging the complexity and uncertainty of this process. The reality is that only a very small proportion of the knowledge generated through research ever reaches a point where it provides a commercial return to the various players in the progression of an invention through to commercialisation (Forfás, 2004a: 16). Consideration of the issues that shape the parameters for commercialisation activities, however, should serve to mitigate some of the risks and help to manage expectations with regard to the likely success of commercialisation activities. This chapter documents macro- and micro-level stimulants to third-stream activities, before considering the institutional, operational and cultural barriers that may constrain technology transfer efforts. Together, these factors determine the nature and extent of technology transfer at HEIs.

The ability of HEIs to build sustainable support for activities and to gain external legitimacy will be largely a function of their ability to demonstrate the impact and benefits of technology transfer. The chapter, therefore, proceeds by discussing the difficulties of capturing and evaluating the impact of technology transfer outcomes, before concluding by providing evidence of measures currently used and highlighting the merits of holistic approaches to evaluating commercialisation and technology transfer activities. The structure of this chapter is indicated in **Figure 5.1**.

Figure 5.1: Structure of the Chapter

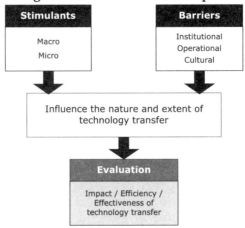

STIMULANTS TO TECHNOLOGY TRANSFER

In order to appreciate fully the potential for commercialisation activities and to place realistic boundaries on expectations, it is necessary to consider the stimulants to technology transfer. These can be grouped as either macro- or micro-level factors.

Macro-level Factors

These include:

♦ **The level of business investment in R&D in the region**: HEIs in regions with higher levels of R&D investment are likely to have greater potential and opportunities to develop links with business and to exploit university research. In some cases, the presence of large multinational pharmaceutical or technological companies in a region will serve to enhance dramatically investment in university research and commercialisation activities.

♦ **HEI proximity to industrial clusters:** Clusters of economic activity often facilitate strong networking between businesses, universities and local government agencies. Furthermore, many of these clusters have developed in technologically-intense industries where R&D and innovation levels are quite high. Over time, student placement and graduate recruitment into clusters should lead to strong and sustainable linkages. These traditional sources of interaction between HEI and enterprises provide a solid foundation for more formalised and strategic linkages.

♦ **Support of development agencies/research councils:** The role and capacity of development agencies to support business-university collaboration at a regional level may prove critical in promoting the growth of third-stream activities. Such agencies may have established linkages with businesses in the region and may also be able to provide first-hand expertise, particularly on the more legislative aspects, of technology transfer and commercialisation. Some countries allow grant recipients to use research grants to pay IP-related costs. For example, research grant applications made by public research organisations to the German BMBF must include an exploitation plan and applicants can include the projected costs in the grant application. The European Union also allows patent costs to be included in the indirect research expenditures eligible for Community Framework grants (OECD, 2003: 45).

♦ **Institutional and legislative context:** Goldfarb & Henrekson (2002) found that, when intellectual property was awarded to universities, it was more effective in facilitating commercialisation than the Swedish system, where the rights are awarded to the inventor. The US legislative context is widely recognised as being the most supportive in terms of providing incentives for encouraging the exploitation of university research. The benefits of this model are that it gives control over IP, provides legal certainty and fosters and encourages technology transfer and public/private partnerships. Further, almost all academic inventions will involve multiple researchers and there is a risk that individual ownership may lead to a fragmentation of property rights, which, in turn, may lead to a hesitation by firms to license technology. For these reasons, the granting of ownership of IP to the research organisation and ensuring that royalties are shared with inventors has emerged as common practice in number of OECD countries (OECD, 2002).

♦ **Government support:** Governments can create a supportive context for technology transfer by sponsoring the establishment of one-stop IP centres or networks to serve the smaller public research organisations that lack the resource or critical mass to build their own fully-operational TTOs – for example, Belgium's Interuniversity Institute for Biotechnology (VIB). In Denmark, as part of the implementation of new legislation, a grant of €8 million was set aside for the period 2000-2003 to help universities protect and market their inventions. The grant was used to help universities cover the external costs of patenting and marketing (up to €20,000 per invention) and to facilitate in the establishment of joint TTOs or research networks, either geographically or by area of research.

♦ **Features of the patenting process:** Often, technology transfer initiatives will be hindered or discouraged by the cost of patent applications. In some cases, macro-level initiatives can ease the financial burden of applications, especially for particular organisations or in particular areas. In the US, for example, the United States Patent & Trademark Office (USPTO) reduced patent fees for smaller entities with fewer than 500 employees. It also lowered patent application fees across the board in 1999, although costs have risen recently (AUTM, 2002). The propensity of academics to explore commercialisation routes will also be influenced by the nature and complexity of the patenting application process. In an effort to address these issues, USPTO has recently implemented an online electronic filing system, while also lowering the average application processing time. Since 1995, USPTO also permits a provisional patent application, which is particularly useful for universities and small firms, as it allows them to obtain early protection for an invention (without preventing the researcher from publishing the results).[16] Attempts at technology transfer are also often impeded poor information and confusion concerning the arrangements for developing and exploiting IP. In the US, clarity was provided through the introduction of legislation under the Bayh-Dole Act, 1980. In an Irish context, implementation of ICSTI's *National Code of Practice for Managing Intellectual Property from Publicly-funded Research* (2004) will go some way in addressing this issue.

Micro-level Factors

This historical context of HEIs will have served to shape their role in the regional/national innovation system and as well as the type of knowledge they generate. Thus, each HEI will have a number of issues specific to itself that will influence the level and extent of technology transfer and research commercialisation.

Micro-level stimulants of third-stream activities may include:

♦ **Research tradition:** Some universities are strongly-grounded in a particular research tradition or, historically, have focused on particular channels of knowledge dissemination. Evidently, more applied subject areas lend themselves to commercialisation activities

---

16 The OECD TTO Report (2003) notes that 'the US patent system allows inventors a "grace period" of 12 months for disclosure, allowing researchers to publish their results and still apply for a patent. Since 1995, the USPTO has offered inventors the option of filing a provisional application for patent. It allows filing without a formal patent claim or an information disclosure statement. It was designed to provide a lower-cost first patent filing in the US and to give US applicants parity with foreign applicants under the GATT Uruguay Round Agreements. A provisional application for a patent has a pendency of 12 months from the date it is filed. The pendency period cannot be extended.'

(Martin, 2000). While previous chapters identified best practice in terms of structure and activities, it is crucial to acknowledge that a well-developed science and technology base, in addition to the availability of a great amount of exploitable research, are *sine qua non* conditions for the successful running of any interface (OECD, 2003; Sanchez & Tejedor, 1995).

♦ **Strategy, mission and clear objectives:** The OECD (2003) notes that effective commercialisation of research requires a specific strategy, in order to ensure the appropriate balance of income generation with the adequate protection of intellectual assets. Furthermore, this strategy should be devised in the context of a HEI's overall mission. Research by Friedman & Silverman (2003) found that a TTO's mission statement was indicative of leadership and an entrepreneurial university culture, which was complementary to a university generating more licenses. Clear objectives stemming from the strategic plan can serve as performance checks. The importance of a clear strategy, sense of purpose and associated leadership is indicated from the case of Cambridge-MIT Institute (CMI) collaboration **Table 5.1**.

♦ **Top-level leadership and commitment:** Universities should co-ordinate the efforts of the various offices that support university researchers in their work with companies and, where appropriate, should consider co-locating them. The university campus president should establish a co-operative tone toward university-industry research collaborations and should align incentives to encourage teamwork and promote research. The encouragement and commitment of top-level management of the university is vital in terms of facilitating implementation of policies, but also so that the TTO and its activities gain legitimacy both internally and to external stakeholders (Meseri & Maital, 2001; Meyer, 2003).

♦ **Quality of the TTO:** Mechanisms for technology transfer will be shaped by the resources, reporting relationships and incentives of technology transfer offices (Bercovitz *et al.*, 2001). Lambert (2003) noted that many businesses report problems with the professionalism of some TTOs and that some universities find it difficult to acquire certain resources such as marketing skills, market research, licence negotiation expertise and spinout experience. Research by Siegel *et al.* (2003) demonstrated the importance of intellectual property attorneys in technology transfer, noting that some universities use these lawyers to help them obtain copyrights and in various aspects of patenting and licensing, especially in support of prosecution, maintenance, litigation and interference. Internal capabilities and access to expertise is a vital stimulant to technology transfer activities.

Table 5.1: Case Study: The Cambridge-MIT Institute

The Cambridge-MIT Institute (CMI) started operations in the summer of 2000 as a joint venture between the two universities. Financed largely by the UK Government with some £65m of grant funding, the objective was to make a dramatic change in the UK's approach to knowledge exchange between universities and business. MIT has an extraordinary reputation as a hub of entrepreneurial activity, and the hope was that its skills could be brought alongside those of one of Britain's great universities to the benefit of the whole economy. With the promise of support for five years, critics say that the project got off to a poor start: its objectives were not set with sufficient rigour, and its internal controls were weak. Earlier this year, it launched a new strategic plan and took on a new leadership. Its life has been extended by a year, at no cost to the public, and it is now working on a range of innovative ideas aimed at improving the effectiveness of knowledge exchange, educating future leaders and developing programmes for change in universities, industry and government.

CMI argues that, without programmes that foster in-depth and interpersonal business university engagement, the contribution such collaborations can make to the economy is likely to be modest. It is now building a series of what it calls knowledge integration communities, which bring together graduates, academics, other universities, companies, suppliers and government agencies to work together from the very start on specific knowledge transfer projects. Examples include an attempt to design a silent aircraft, and research into pervasive computing and nanotechnology. CMI's success will be judged by the sustainability of its programmes at the end of the six-year period, by its success in building a bridge between two of the world's great universities and – above all – by its ability to develop new types of partnership between businesses and universities across the UK to the benefit of the whole economy.

Source: Adapted from Lambert (2003).

♦ **Age and experience of TTOs**: Developing expertise in commercialisation activities can take a significant amount of time. The AUTM (2002: 7) notes: 'this is a significant factor in comparing performance because of the time needed to develop a portfolio of intellectual property to license, build up a body of expertise and develop a culture of technology transfer within the institution – as well as giving licenses the time needed to develop and market products'. The development of expertise and tacit knowledge in commercialisation over time, therefore, is a necessary condition for success (Allan, 2003; Hernes & Martin, 2000; Stevens, 2003; Molas-Gollart *et al.*, 2002; OECD, 2003; Siegel *et al.*, 2003). Business will tend to deal with an older TTO that has built up a reputation, as it will provide more experienced professionals who have dealt with issues in the past (BHEF, 2001). These factors go some way in explaining the advantage that the USA has in this area (see **Figure 5.2**).

Figure 5.2: Technology Transfer Programme Start Dates of US Universities

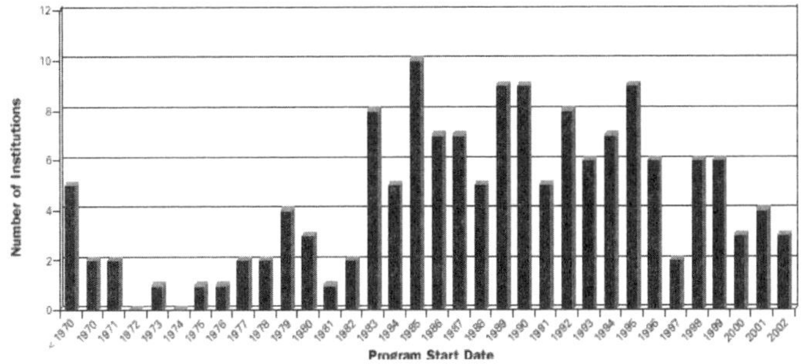

Source: AUTM (2002).

♦ **Scope of commercialisation initiatives:** A number of institutions are implementing a variety of means to foster an entrepreneurial environment and encourage relationship management (Meyer, 2003). Traditionally, it has always been recognised that the most effective forms of transfer involve human interaction as opposed to arms-length transactions. These relationship management programmes are often directed far beyond inventors or likely inventors. Outreach to the community in which the university resides, particularly in the case of state-funded institutions, is on the rise. This trend is more apparent at universities not ordinarily considered in the top tier of schools as ranked by criteria typically measured by AUTM surveys. For example, one university studied by Scott *et al.* (2001) hired an outside firm to measure customer satisfaction with its TTO. Several TTOs in the OECD survey reported holding annual recognition events, such as banquets or luncheons, for inventors who have been granted patents during that year (OECD, 2003).

♦ **Documented policies:** Clear and well-documented, written policies pertaining to most aspects of technology transfer are of real benefit in encouraging commercialisation activities. The extent to which universities have established clear and detailed rules on property rights to new ideas or inventions show great variation (Hernes & Martin, 2000). Such documents are often available on the Internet or internal university intranet. Increasingly, efforts are made to make these documents as user-friendly as possible, so that the process of technology transfer can be understood by researchers. More advanced universities are now providing summaries, as well as pocket-guides, to ensure that all TTO stakeholders are aware of, and adhere to, policies. In the absence of clear policies, as technology transfer activities expand, the situation becomes much less tidy.

Tensions are generated between university central authorities and academic entrepreneurs in academic departments and research centres, and questions of co-ordination, adherence to university regulations become issues of great sensitivity (Shattock, 2001).

♦ **Educational offerings:** Many TTOs are now offering workshops and presentations to increase awareness of technology transfer and to enhance understanding of the processes in place in their respective institutions. Courses provided by the AUTM are useful benchmarks, demonstrating a strong lead to facilitate this learning (Allan, 2001; OECD, 2003).

♦ **Knowledge of research and relationships with faculty:** Motivating and helping researchers to locate potential collaboration partners requires a sophisticated understanding, not only of how researchers operate but also of individual researchers' focus areas, and of the companies that share their research interests. When technology-transfer, sponsored programmes, or corporate relations officials are knowledgeable about faculty research interests, they can play a key role in pre-screening companies with which the faculty might wish to collaborate. Deans, department chairs, and vice presidents of research are well-positioned to co-ordinate these efforts. According to the OECD, 'close relationships with inventors and labs are necessary for technology transfer' (2003: 17).

♦ **Networking and informal relationships:** Most TTOs surveyed by the OECD (2003) stressed the importance of *both* the TTO's and the researchers' informal relations. This may suggest why, in many cases, it is recommended that the staff hired for TTOs have industry experience and, hence, an understanding of industry but, equally important, a network of contacts and linkages (OECD, 2003). Success stems from recognising the importance of such organic informal mechanisms and attempting to complement them, rather than cannibalise them, with new initiatives (see Stevens, 2003).

♦ **Trust and common expectations:** A minimum level of trust is a pre-condition for effective technology transfer (Lundvall, 2002: 8). To achieve successful co-operation agreements, both parties need to be aware of each others' interests and objectives as well as each others' complementary strengths (Fassin, 2000: 32). An understanding of various stakeholder objectives (Siegel *et al.*, 2003: Chapter 3), and developing consensus in terms of expectations through communication and negotiation, is critical to technology transfer initiatives and to minimise the potential conflicts of interests that may arise.

♦ **Inventor involvement:** Increasingly, it is acknowledged that transfer of knowledge from the university to the commercial sector generally requires the active involvement of university inventors (Goldfarb &

Henrekson, 2002; Meseri & Maital, 2001). In the US, for example, it has been noted that ideas reach TTOs in primitive states and much critical knowledge is often tacit. Jensen & Thursby (2001) found that at least 71% of inventions require further involvement by the academic researcher if they are to be successfully commercialised. Some 48% of the ideas are in proof of concept stage, or laboratory scale prototypes (29%) and for only 8% is manufacturing feasibility known. A stimulant to successful technology transfer, therefore, is knowledge and know-how at both the sending and receiving ends. As captured by the seminal work of Burns & Stalker, 'the design, which is the outcome, is as much determined by the effectiveness of the relationship between technologist and the user as by the individual contributors on either side' (1961: 40).

♦ **Motivation, incentives for researchers and university culture:** [17] These are a major stimulus and critical input to technology transfer (Siegel *et al.*, 2003). Faculty involvement or buy-in to technology transfer is crucial, as one cannot assume that invention disclosures will be made; these must be encouraged and the validity of this process for research findings actively promoted. Researchers who have experienced academic entrepreneurship elsewhere display a greater propensity to exploit their intellectual property for commercialisation purposes. European academics have not shown the same enthusiasm for research commercialisation initiatives as witnessed in the US. One explanatory factor for this is the overall level of priority which the academic system attaches to links with industry and to the practical application of research results. Siegel *et al.* (2003) found a strong relation between the university-based 'supply' of technologies for commercialisation and the university's tenure, royalty and distribution policies. Participation in commercialisation activities should be regarded as a valid and legitimate contribution towards aspects of the university mission. As noted by Hank McKinnell, Chairman and CEO, Pfizer Inc., 'Technology transfer is people to people. You have to commit the people to make it work'. Yet, while incentives are useful in creating an atmosphere conducive to technology transfer, ultimately efforts must be based on a voluntary willingness by researchers to exploit and disseminate their research findings. This is best done through creating a culture in the university that values and encourages innovation and entrepreneurship (Den Hertog *et al.*, 2003).

♦ **Realistic expectations:** In an effort to promote the value and potential impact of commercialisation activities, it is easy to be consumed by rhetoric. It is worth remembering, however, that

---

[17]   The importance of these issues is a recurring theme throughout this book, as summarised in **Chapter 8**.

although a very few, and highly visible, blockbuster inventions, such as the Cohen-Boyer gene-splicing patent from Stanford University, have made tens of millions for universities, most university licensing offices barely break-even (Allan, 2001). This reality may need to be revisited occasionally, to reign in overly-optimistic views of politicians and high-level administrators. Nelson (2001) amplifies this conclusion: 'The direct economic impact of technology licensing on the universities themselves has been relatively small (a surprise to many who believed that royalties could compensate for declining federal support of research)'. Given the time lags involved, technology transfer requires patience and persistence (Meyer, 2003). Typically, there is a four-to-nine-year lag between making a discovery until the first introduction of a new commercial product or process based on that discovery (Mansfield, 1991). Furthermore, these effects are often complex and difficult to attribute causality to (for example, development of curricula or transfer of tacit knowledge). The purpose of this cautionary note is not to introduce scepticism, as TTO activities should not measured only by their narrow economic impact but rather by a multitude of variables. Instead, there should be realistic expectations of the financial benefit of commercialisation activities, coupled with an appreciation of the broader impact of ILO activities in terms of facilitating the achievement of the university's mission.

BARRIERS TO COMMERCIALISATION

Van Dierdonck & Debackere (1988) identify three barriers to commercialisation: institutional barriers (unclear norms and policies, resource and expertise constraints); operational barriers (constraints on research, motivational issues); and cultural barriers (mutual incomprehension). Using these barriers as a structural framework, this section documents the major barriers for commercialisation as depicted by the literature and empirical research in this area.

Institutional Barriers to Technology Transfer

These include:

♦ **TTO organisational structure and supporting infrastructure**: Typically, university structures have not been designed to carry out technology transfer. Often, structures have evolved in an *ad hoc* manner and are not guided by any specific policy or clear divisions of labour/job descriptions, so that the contact point and process of commercialisation is not clear for academics. Historically, TTOs have not been afforded a prominent position in HEIs' organisational structure or among HEI management teams. Further, as a result of historical development, work is often overlapped between university administration offices. In this context, it is rare for TTOs to leverage their knowledge for mutual benefit. Consequently, the process of commercialisation can be impeded by bureaucracy and poor communication between various parties. Often, attempts to rectify these problems, by moving TTOs and Research Offices closer, for example, are restricted by lack of available property or building constraints.

♦ **Resource constraints and expertise deficiencies:** In the past, commercialisation activities have not received adequate attention from university top management. As a result, traditionally, they have been under-funded and under-staffed. Recent developments have brought about changes in this area but, as a result of rapidly evolving technologies and legislative developments, often TTOs do not have sufficient expertise, particularly in terms of knowledge of the patenting process. TTO staff also frequently lack the business and marketing skills to lead the commercialisation process pro-actively from invention disclosure to patenting and licensing. Thus, very rarely do TTOs adequately fulfil the role of 'science marketer'. This can prove detrimental in the long term, as universities can have filed and received patents for which they are receiving no benefits or income, since they have not been licensed to business. Further, many TTOs lack sufficient resources, particularly in covering the cost of patent registration. The cost of registering an Irish patent is prohibitive for some TTOs and is a barrier to eventual commercialisation. Such difficulties are accentuated by the fact that the cost of obtaining a patent in Europe is up to five times higher than in the US (Connelan, 2004). The impact of such expertise and resource deficiencies becomes most obvious when it comes to managing spin-offs and start-ups. The OECD *Turning Business into Science* report (2003) noted that the limited success in this area can often be attributed to the lack of sound management of technology transfer, as well as insufficient IP and legislative know-how.

♦ **Poor internal relations:** Poor experiences in dealing with the TTO, or long time delays in terms of feedback on invention disclosures, can reduce the propensity of academics to use the resources of the TTO. Scott *et al.* (2001) notes that bad relations between TTO staff and faculty/academics can be a huge barrier to commercialisation.

♦ **Lack of clarity over ownership of intellectual property**: Lack of clarity on IPR can make negotiations longer and more expensive than would otherwise be the case and, in extreme cases, can prevent them from being completed (Lambert, 2003; Sanchez & Tejedor, 1995). Barriers are raised by complex proprietary concerns, especially in the case of jointly-sponsored research contracts. The approaches of some companies and other funding entities in the area of IPR, such as the pursuit of extensive rights to university background research developed outside the collaboration, can prove to be stumbling blocks. Research by Siegel *et al.* (2003) found that the private sector expresses frustration with obstacles that impede the process of commercialisation, such as disputes that arise within the university regarding intellectual property rights.

♦ **Perceived conflicts of interests:** Perceptions of a conflict of interest can damage the TTO and institution by weakening public trust. Institutional conflicts of interest, also called 'conflicts of mission', are a newly-emerging source of concern. Some universities invest in start-up firms, or accept equity in lieu of royalties on university-held patents, raising concerns that they might become beholden to a company in which they have a financial stake. Thus, TTOs need to be careful about the commercialisation activities that they conduct and their resultant impact, in terms of external perceptions of the TTO and the university (BHEF, 2001).

♦ **Overestimating the value of intellectual property:** There is a danger that some HEI leaders might consider technology transfer activities solely as a revenue source, as opposed to a component of the university's mission and overall public responsibility. Such attitudes can raise barriers to negotiations, and actually reduce revenue over the long-term. Premature definition and valuation of intellectual property can become an obstacle at the initiation stage of a collaborative project. Attempts at maximisation can contradict the mission of the TTO, while, at the same time, acting to the detriment of the commercialisation process (Sanchez & Tejedor, 1995).

♦ **Patenting of research tools may discourage beneficial research:** This topic increasingly appears to be an issue in the biomedical area. For example, different universities may hold patents to different receptor cells of the same class that influence a disease process (see OECD, 2003, p.64).

♦ **Partners may lack understanding or trust:** In some cases, partners enter into agreements with an inadequate understanding of the management, internal politics, decision-making structures, and even fundamental interests of the other partner, resulting in slow decisions and insufficient resources. Of the many ingredients in a successful negotiation between companies and universities, mutual trust is perhaps the most important.

♦ **Lack of support for SME technology transfer:** Although most of the firms in a region are likely to be SMEs, very few TTOs provide specific contact points for these types of organisations. Often, such firms do not have the personnel, time or financial resources to invest in technology transfer activities and have very limited knowledge of them. For TTOs, this makes contact and negotiations with SMEs challenging and time-consuming. However, for those universities that do provide specific SME-related contact points, there can be huge benefits.

♦ **Complicated technology transfer policies:** One barrier to commercialisation that reduces the propensity of an individual researcher to disclose an invention is the practical difficulty of negotiation and the complex and burdensome nature of the commercialisation process. If policies are not clear or user-friendly, staff will be reluctant to follow the commercialisation route, particularly given their time and knowledge constraints.

Operational Barriers to Research Commercialisation: Constraints & Activities

These include:

♦ **Lack of space:** In most instances, pursuing new research activities requires the recruitment of new researchers, the acquisition of additional physical space and access to laboratory equipment. The current pressure on physical space and other resources in most HEIs, however, means that these resources are not readily available. This adds another barrier to the successful completion of the research goals and possible commercialisation opportunities.

♦ **Lack of investment in R&D by companies:** A criticism frequently levelled at industry is that it does not invest sufficient resources in long-term research and development, despite policy initiatives to encourage such investment – for example, the *Lisbon Agenda*. The reality is that investment undertaken by companies is largely devoted to overcoming short-term operational issues and focused on cost reduction. This continues to be a significant barrier to the commercialisation of research, as it cuts research off at its roots.

♦ **Lack of funding for prototype development:** There also tends to be a funding gap between research and prototype development, which is a major barrier to commercialisation. When the period of a research grant ends, and the project is at a stage where it requires assistance to take it to the next step to make it commercially-attractive to companies, in most cases, there is no funding available for this activity. Consequently, pressures to attract new research funding force researchers into applying to new agencies and commencing entirely new, possibly unrelated, projects.

♦ **The 'research treadmill':** To maintain a reasonable research programme with research staff and post-graduate researchers, senior researchers must constantly submit applications for additional research funding. This has been referred to as the 'research treadmill' (Pandya & Cunningham, 2000). A difficulty occurs where there are no funds available for researchers to develop their existing innovations further and they must submit proposals for new basic research, when they have not taken previous research to a stage where it is ready for commercialisation or technology transfer. Additionally, the constant requirement to win research funding is a time-consuming process, taking away much of the time that researchers otherwise might use for commercialisation or technology transfer activities.

♦ **Availability of researchers to pursue research fellowships:** In the case of many research projects undertaken, postgraduate researchers form an essential part of research project teams. In particular areas, the recruitment of high-calibre individuals can prove difficult. This is particularly the case in emerging areas, where attractive opportunities often open up to graduates outside the research arena. In Ireland, this deficiency has been resolved through the recruitment of foreign students. Though beneficial, the weakness of such an approach is the leakage of intellectual property should students return to their home countries.

♦ **Constraints for researchers:** On an individual level, academic staff have little time either to establish or to undertake collaborative projects with industry, in addition to their teaching and administrative duties for the HEI. Moreover, the continued emphasis on traditional outputs for academic work, such as publications, has meant that collaborative industrial R&D is not valued, except as a source of income. Furthermore, the general lack of academic recognition for commercialisation and, in contrast, the rewards for publications, as opposed to patents, has been a major barrier to the commercialisation of research (BHEF, 2001). As a result, many academics have been faced with the dilemma of either publishing their results for short-term revenue and academic recognition or withholding them until they are patented, with the risk of technology becoming obsolete. Other factors limiting the motivational drive of academics include:

  ◊ **Perceived publishing constraints:** Many academics feel that publication delays and non-disclosure requirements may impair the openness of the university research environment and subsequently affect promotion and tenure decisions. This concern is believed to be particularly strong in the case of younger academics, as delays may impair them from building the strong record of publications needed to gain tenure.

◊ **Age of academic:** The age of the academic is often an issue in terms of the likelihood of commercialisation. Some argue that younger academics are more likely to commercialise, as they are less risk-averse than their older colleagues. Others suggest that older academics are more likely to undertake commercialisation activities, as they have more experience and may not be so 'career-orientated'.

♦ **Contract researchers and salaries:** There is an increasing tendency at institutions to employ a high number of research scientists on short to medium-term contracts. Consequently, due to the lack of permanent staff, these institutions are hindered in their efforts to develop and retain a critical mass of expertise in key research areas. A parallel issue resulting from the high turnover of contract staff is the disappearance from the technology cycle of the individual researcher with greatest knowledge and understanding of the technology, before the technology has reached commercialisation stage. Given the current ceiling on salaries, HEIs are somewhat constrained in their ability to attract and retaining specialist academic staff.

♦ **Lack of full-time research positions:** There is no tradition of pure research careers in Irish HEIs. Once an individual has completed post-doctoral studies, generally the only option for them to remain in the institution is as a lecturer. This serves as a barrier to the further development of research in commercial terms by the researcher involved in the project and most familiar with the technology. Once the researcher has left the project, even to lecture in the same department, time constraints make the pursuit of research less likely.

♦ **Confidence in commercial merits of the research:** Some academics do not appreciate, or are unable to estimate, the commercial value of research they have undertaken. Hence, they may not be motivated to exploit the commercial potential of the project.

Cultural Barriers: Mutual Incomprehension

Business and HEIs are not natural partners: their cultures and their missions are different. Academics value their freedom and independence, resent their reliance on public funding and feel their efforts are not properly appreciated (Lambert, 2003). Reflecting this, the most frequently-cited barrier to commercialisation, both in the literature and from our research, was the cultural barriers that exist between the TTO, the university scientists and industry. Naturally, different stakeholders have differing expectations but, in the university context, such differences are magnified. Research by Sanchez & Tejedor (1995) made reference to managers who rated their relationship with university staff as of little benefit because of the 'the impeding impact of a cultural barrier'. **Table 5.2** attempts to capture the main differences identified through our research.

Table 5.2, while simplifying differences, clearly indicates how collaboration between the two parties may prove difficult. Universities, for example, are dedicated to academic freedom, publication and self-determination (Graff *et al.*, 2001). In this respect, faculty members tend to be concerned with tenure and promotion decisions, as well as salary increases based on merit. Business, on the other hand, is driven by a clear product-driven focus and a culture that emphasises applied research, secrecy, and protection through patents (Fassin, 2001; Nelson, 2001). The focus tends to be on short-term results and targets, particularly in sectors such as the food sector. In addition, some sections of industry view universities as a low-cost route to solve immediate operational problems. Research by Jones-Evans (1999), comparing ILOs in Sweden and Ireland, similarly found cultural differences manifest in different conceptions in terms of time, priorities and bureaucracy.

Table 5.2: Cultural Differences: Universities *vs* Industry[18]

University Values	Industry Values
New invention	New application
Advancement of knowledge	Added value
New means for further research	Financial returns
Basic research	Applied
Long-term	Short-term
To know how? What? Why?	Product-/service-driven
Free public good	Secrecy
Publication	Protection/patents
Academic freedom	Commercial approach
Supply-side model of action	Demand-side model of action

Technology transfer will only occur when university faculty and representatives from business and industry work together for mutual gain and find mechanisms to manage the inherent conflict between openness, characteristics of the scientific community and the privacy/secrecy that belongs to the world of business. Therefore, industry-university collaboration cannot be forced and cultural

---

[18]   Obviously, such extreme polarisation glosses over issues that are remarkably diverse and complex – for example, discussion of basic *vs* applied research (see **Chapter 1**).

differences must be understood. More specifically, particular commercialisation barriers arise from the following:

♦ **Departmental and faculty attitudes towards commercialisation:** In some university departments and faculties, there can be an attitude that commercialisation is not a suitable activity for academics. This view considers that the research undertaken by individual academics should be available for the public good and is not to be exploited for a return for the academic or their institution. This is a significant barrier to commercialisation. In some instances, this view is held by established academics who hold key positions such as Head of Department or Dean and who thus play a lead role in the creation of a culture where commercialisation by academics is not valued, encouraged or supported at departmental, faculty or institutional levels. Academics who proceed to commercialise their intellectual property rights are often regarded as mavericks and normally suffer the ultimate sanction of finding it difficult to get promoted. Furthermore, the research training and background of some permanent academic staff means they have never experienced academic entrepreneurship in operation. Consequently, their mindsets have not broadened to accept the value of commercialisation undertaken by younger academic staff.

♦ **Lack of awareness:** Many academics have little knowledge or understanding of the patenting process. The general understanding is that a patent protects one's ideas; however, in practice, the patent is very much a lateral application. Therefore, there is an awareness gap in terms of understanding patents and determining their different uses. Furthermore, it is often the situation that the academic is unaware that they are unable to publish directly-related material directly prior to filing for a patent. In addition, there is a lack of awareness regarding the length of time before patents lapse. This lack of understanding and awareness also extends to Technology Transfer Officers, who in many instances do not have the expertise to advise academics through the complexity of patent filing. In a similar vein, Forfás (2004a: 26) attributes the low level of patenting by public research organisations to the following:

◊ Low levels of awareness of IP among institutional researchers, particularly in the Institutes of Technology.

◊ Shortfall in the availability of staff with expertise in IP management.

◊ Low levels of commitment from the universities and Institutes of Technology to producing patents.

◊ Low levels of funding for patents and associated activities.

♦ **Perceived non-challenging nature of commercialisation:** Some respondents expressed the opinion that academics do not perceive commercialisation as intellectually challenging and that they are not excited to see their research being commercialised. This may reflect the historical viewpoint of a department, and can be overcome only with education and communication as to the benefits of commercialisation for the public good.

♦ **Lack of soft skills:** A deficiency issue in soft skills can also form a barrier to commercialisation activities. These soft skills include communication skills, presentation skills, time management and the ability to negotiate. These skills particularly come into play when a researcher is undertaking research within industry and also in the adoption phase by industry.

Barriers to Commercialisation: Empirical Evidence

While the barriers identified in this chapter have been drawn from various studies and literature in this area, in order to illustrate the differing impact of barriers from different stakeholder perspectives the work of Siegel *et al.* (2003) is insightful. Siegel and his colleagues conducted 55 structured, face-to-face interviews with UITT stakeholders (15 administrators, 20 managers or entrepreneurs, and 20 scientists affiliated at five universities) and complemented this qualitative data with statistical analysis from the AUTM survey. The results in respect of three stakeholders, namely scientists, managers and TTO/ILO managers are shown in **Figures 5.3**, **5.4** and **5.5**.

Figure 5.3: Barriers Identified by University Scientists

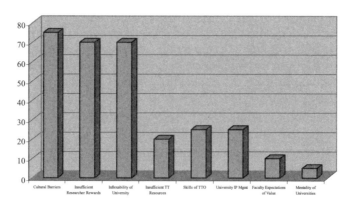

Figure 5.4: Barriers Identified by Managers

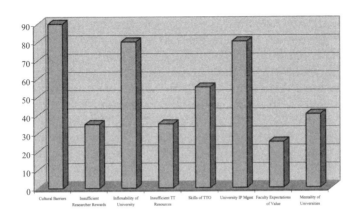

Figure 5.5: Barriers Identified by TTO Managers

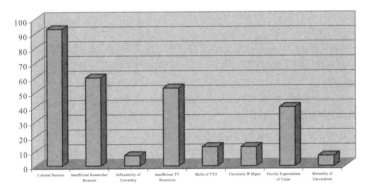

EVALUATING TECHNOLOGY TRANSFER ACTIVITIES

A major barrier to legitimising and gaining buy-in for commercialisation activities stems from difficulties in identifying appropriate activity and outcome indicators. Currently, Ireland does not have activity indicators or reporting requirements in place to measure managed technology transfer and progress in the commercialization of R&D (Forfás, 2004a: 35). Clearly, this is a deficiency, although it reflects a general difficulty in that the national systems of innovation, which underlie and help to explain performances, involve complex phenomena that are difficult to measure and analyse. Typically, the most commonly used measure of technological output is patenting activity (Forfás, 2004).

The Difficulty of Measuring TTO Activities

This book seeks to illustrate the wide range of technology transfer activities and mechanisms available for the commercialisation of research. Throughout, the emphasis is to acknowledge the complex and uncertain nature of these attempts at fostering innovation. This complexity necessarily makes attempts to measure the impact of technology transfer activities difficult, as there is a potential wide scope of benefits, many of which are complex and arise through indirect channels, which make causality difficult to attribute (Mowery *et al.*, 1992). Therefore, the true impact of TTO interventions on the technology transfer process is difficult to measure quantitatively (Fassin, 2000: 40). This difficulty is accentuated both by the heterogeneity in TTO objectives and by multiple stakeholder involvement, which can mean that success is often the result of complex relationships which involve the whole scientific community (Thursby *et al.*, 2001).

Despite these difficulties, there is a requirement to measure the activities and progress of TTO development. If these measurements are linked to university-wide objectives, they can promote co-operation and a sense of common purpose. Extreme caution, however, must be exercised against measurements that focus exclusively on the revenue-generating activities of technology transfer as this will perpetuate:

♦ The negative image of commercialisation activities among already cynical staff.

♦ The idea of technology transfer as a major source of revenue generation for HEIs. The true revenue-generation effects of technology transfer, even in the US, however, are extremely over-estimated and, in reality, quite minimal (BHEF, 2001).

Royalty income, for example, is an important measure of technology transfer performance, because of its direct relation to economic impact. Yet, in the context of total sponsored research funding, royalty income is a relatively minor contribution to the finances of institutions. In the US, it is as little as 3% of total institution finances and is thus easily dwarfed by tuition income and charitable donations (Stevens, 2003).

The introduction of indicators cannot be seen to be proposed merely for their role as performance metrics but rather must be promoted to staff for the way in which they will facilitate university administrators in the support and management of activities. Crucially, there must be buy-in from TTO staff, as such indicators must be useful to those who will be responsible for their collection, as opposed to being viewed as an external imposition for the sole purpose of measuring and rewarding past performance (Molas-Gollart, 2002: 46).

In order to have utility for TTO managers, evaluative indictors must therefore be SMART: **S**imple, **M**easurable, **A**ctionable, **R**elevant reliable and reproducible and **T**imely. Ultimately, following these guidelines will ensure that indicators are transparent and easily communicated, as opposed to the theoretically-complex models that tend to prove difficult to operationalise. In order to develop such indicators, a number of challenges must be recognised and addressed.

Challenges in Developing Metrics to Measure & Evaluate TTO Activities

These include:

♦ **Need for 'local' metrics:** There is no one model of the successful university. Each university is a product of a distinct process of social, economic and intellectual development. Developing a set of indicators will require a degree of flexibility in how the indicators and their emphasis are applied to different types of university. The system of indicators should allow for a 'variety of excellence' to emerge (Molas-Gollart *et al.*, 2002).

♦ **Organic nature of university-industry interactions:** The relationships between university and industry are often subtle, informal and linked to personal exchanges between individuals. Past studies have found that many of these interactions are often immune from direct influence by policy or management interventions (Mowery *et al.*, 2001). Traditionally, universities have found it difficult to encourage staff to document such interactions and, therefore, the full extent of university-business interaction, or the benefits stemming from this interaction, can prove difficult to measure. This organic nature of interaction limits the ability to find suitable measures and instruments to shape behaviour. It is possible that there will be no one-to-one matching between policy actions and interaction patterns (Molas-Gollart *et al.*, 2002). Research by Sanchez & Tejedor (1995) has indicated that links are often established informally, without the assistance of a liaison office.

♦ **Differences across disciplines:** The level, extent and nature of interaction and commercialisation mechanisms will differ by discipline. Naturally, there will be differences between applied disciplines (mechanical engineering, business administration or medicine) and fundamental theoretical disciplines (theoretical physics or philosophy). In the former, direct channels of application may exist; in the latter, the impact of academic activities on the economy and social welfare is likely to be more long-term and indirect. Measurement of technology transfer will need to be all encompassing in order to capture all these activities and effects.

◆ **Resistance/lack of enthusiasm for evaluative metrics:** There is little enthusiasm in universities for new measurement systems. Often, such systems can be seen as a resource constraint and administrative burden, or staff feel as if such metrics should be reserved for private sector organisations. Promotion of the benefits and objectives in setting up such metrics is therefore critical.

◆ **Inter-dependence between university activities:** These will only be captured by a university-wide measurement system. The discussion on organisational structures (see **Chapter 2**) highlighted the varying implications of organisational structures, in terms of aligning incentives and control mechanisms. At John Hopkins University, for example, there are unit-level incentives, and unit-level information processing capacity. It is sometimes the case, however, that sponsored research revenue is leveraged against licensing revenue, and this may not be captured by traditional measurement systems.

◆ **Indirect impacts:** Related to the above, technology transfer activities may also result indirectly in research contracts, grants and donations to the university, as well as an enhanced student intake as a result of reputation/linkages with companies (Graff *et al.*, 2001).

◆ **Serendipity:** The outcome, and therefore the impact, of research activities are by their very nature unpredictable, and serendipity is an important element when companies are attempting to develop new products and generate competitive advantage.

◆ **Efficiency *vs* effectiveness:** It is often the case that measurement systems focus on quantitative output and measurement can be reduced to looking at numbers – for example, invention disclosures, patents applied for, patents granted. While useful, such metrics do not capture the effectiveness of the process (for example, the extent and nature of interaction between TTO staff and academics), or the quality of the activities conducted (for example, evaluating an invention disclosure in terms of its patentability).

MEASURING TECHNOLOGY TRANSFER ACTIVITIES: OUTPUT & IMPACT INDICATORS

Association of University Technology Managers

Perhaps the best known survey of commercialisation activities is the annual evaluation undertaken by the Association of University Technology Managers (AUTM) in the US. The aim of the survey, which began publication in 1993, is to monitor the patenting and licensing activity of US and Canadian universities, research institutes and teaching hospitals in order to show how they are making federally-funded inventions available to the public. Much of the evidence showing increasing levels of commercialisation of university research in the US is based on the results of this survey. The AUTM approach represents good practice in the elaboration and implementation of surveys oriented to the assessment of universities' commercial activity. It provides a structured approach with firm definitions, which have been refined over time. AUTM measures focus on quantitative metrics for technology commercialisation and licensing, such as numbers of patents, licenses, licensed products, commercialised products, and income generated.

The limitations of the AUTM survey are that the instrument is very much tailored for the US context and, therefore, requires adaptation for application in other contexts. Results are often subject to 'halo effects' – after a university has gained visibility as a result of its success, it tends to receive additional attention just because of such visibility. Further, data, particularly in terms of revenue income, is often heavily skewed as a result of one extremely successful commercialisation, so that it is difficult to conduct an overall evaluation of the HEIs technology transfer activities. Patent counts have the disadvantage that they only take into account the quantity of patents, while neglecting the fact that patents can differ dramatically in quality (Coupe, 2003: 32). Coupe suggests a more robust way to measure the quality of patents, by using *subsequent* patent citation counts. Nevertheless, the AUTM survey provides a useful template for looking at TTO activities. Sample questions drawn from the survey in this respect are highlighted in **Table 5.3**.

Table 5.3: Quantitative Measures for Technology Transfer

Sample Questions from the AUTM Survey (2002)

1. Research expenditure:
◊ Total research expenditures.

2. License/option agreements:
◊ No. of licenses executed (exclusive/non-exclusive).
◊ No. of licenses/options that included equity.
◊ No. of licenses/options active on last day of fiscal year.
◊ No. of licenses/options executed in fiscal year licensed to start-ups.

3. Research funding (related to licenses/options):
◊ Research funding committed to institution in fiscal year related to license/option agreements.

4. License income:
◊ No. of licenses/options yielding license income in fiscal year.
◊ No. of licenses yielding running royalties.
◊ Total license income of the institution (% attributed to running royalties, cashed-in equity, license income of all other types).

5. Legal fees, expenditures and reimbursements:
◊ Amount spent on external legal fees for patents and/or copyrights.
◊ Amount received in reimbursements for these fees from licensees.

6. Patent-related activity:
◊ No. of invention disclosures received.
◊ Total (US) patent applications filed.
◊ No. of (US) patents issued.

7. Start-up companies:
◊ No. of start-up companies formed during fiscal year, dependent upon the licensing of your institution's technology for initiation.
◊ Of the start-up companies formed during fiscal year, in how many does your institution hold equity.

8. Licensed technologies, post-licensing activity:
◊ Did any of your institution's licensed technologies become available for consumer or commercial use during fiscal year.

OECD Studies

The OECD has launched an initiative to examine how the strategic use of intellectual property rights is evolving in public research organisations. The most recent manifestation of this was the *Turning Business into Science* survey (2003). The project aims to establish international comparability of data by suggesting a standardised methodology and some core questions to be included in future questionnaires. This initiative considers data such as:

♦ Number of TTOs and TLOs per research university.

♦ Funds committed to IP management – either by PROs and universities, or through national programmes.

♦ Number of patents (either patent applications or grants) by universities and PROs.

♦ Licensing and royalty revenues.

♦ Types of technologies that are patented and licensed.

♦ Number and size of research contracts with the private sector.

♦ The cost and frequency of litigation over infringement of intellectual property rights.

Benchmarking

Another approach to measuring and evaluating technology transfer activities is to benchmark quantifiable measures against best practice institutions. As an example of this, the key technology transfer indicator values for a number of top US universities for the year end 2002 are shown in **Table 5.4**.

The problem with such simplistic comparisons is that they are acontextual and take no account of institutional factors, resources available for technology transfer, or alternative routes for commercialisation. In an Irish context, a better comparison may be with UK institutions, the key measures for which are shown also in **Table 5.4**.

Coupled with the difficulty of directly comparing institutions, best practice in technology transfer is not easily identified; a multitude of variables impact on the possible outcomes of a technology transfer project. Graff *et al.* (2002) note that 'universities and their TTOs are still climbing a learning curve of best practices in patent protection and licensing'. Case study considerations of best practice, however, suggest a number of commonalities among best practice institutions (see also **Chapter 6**). Typically, there is:

♦ **Clarity** in the university's policy and procedures for exploitation of intellectual property and commercialisation.

♦ **Efficiency** of internal structures and processes in delivery of technology transfer and commercialisation outcomes.

♦ **Synergy** of technology transfer and academic personnel, with emphasis on access to information and contacts.

Table 5.4: Key Technology Transfer Indicators

Rank	Institution	Total research	Disclosures	Patents filed	Patents issued	Licenses & options executed	Licences & options yielding	Gross license income	Start-ups
1	MIT*	$727.60m	425	316	152	102	362	$30.23m	31
2	Stanford University*	$444.27m	252	240	98	162	378	$34.60m	8
3	North Carolina State University	$415.62m	169	103	45	47	76	$2.56m	6
4	Columbia University*	$311.12m	194	108	78	46	143	$138.56m	7
5	Ohio State University	$289.48m	106	61	28	29	35	$1.81m	2
	Median US	**$294m**	**113**	**74**	**28**	**41**	**65**	**$3.60m**	**4**
	Mean, 74 UK Universities	**N.A.**	**30**	**11**	**9**	**6**	**11**	**Stg£433k**	**2**

*Private Institutions

Source: University of Arizona Office of Economic Development, Technology Transfer at the University of Arizona: A Comparative Analysis and Benchmarking Study, May 2004, based on Association of University Technology Managers (AUTM), Licensing Survey FY 2002; Compiled from Annual Survey of University Technology Managers.

Additional Studies

In his historical analysis of descriptive statistics and indicators in this area, *Twenty Years of Academic Licensing: Royalty Income & Economic Impact*, Stevenson (2003) identified a number of measures that had been set to measure technology transfer (see **Table 5.5**).

Table 5.5: Measuring Technology Transfer

Measurements used in previous research, as identified by Stevenson (2003), include:

- Total R&D funding.
- Total industrial funding of R&D.
- No. of invention disclosures.
- No. of patents pending.
- No. of patent applications filed.
- No. of copyright applications filed for software.
- No. of inventions licensed.
- No. of licenses issued.
- No. of licensed technologies earning royalties.
- Gross royalty income.
- No. of start-up companies formed.
- Staffing levels.
- Programme cost.

A recent empirical investigation by Friedman & Silberman (2003) used output variables, such as number of licenses, number of licenses with equity, number of start-up companies, and total royalty income as output measures. Additionally, some TTO/TLO and academic studies have collected complementary information on patenting and licensing activities and entrepreneurship. This information has been used to develop more complex indicators and statistical inferences that help TTO/TLOs to manage their activities and prove their economic contribution to society. The data and estimates that have been generated and analysed include (Allan, 2001; Molas-Gollart *et al.*, 2002):

- The value of pre-production investment undertaken by firms involved in licensing university technology.
- Estimates of the jobs generated by firms involved in licensing university technology and attributable to the new activities launched due to such licences.
- Estimates of taxes generated by these activities.
- Contributions to local businesses.

While extremely important economic indicators, the limitation of such broader 'impact studies' is the fact they draw on complex theories and methodologies that are difficult to communicate and replicate. Furthermore, because they are often US-based, they assume a scale and scope in TTO activities that may not be present universally. Also, efforts tend to neglect the fact that 'it often takes a considerable time before new scientific knowledge has a tangible economic effect' (Scott *et al.*, 2001). Licensing agreements, for example, vary substantially in their significance, making it dangerous to draw inferences about aggregate technology flows based on the number of deals. Figures on licensing revenue also can be a somewhat misleading indicator of the current performance of a TTO, as the royalty streams may reflect transactions that were consummated many years ago. Drawing therefore on the recent work of Molas-Gollart *et al.* (2002), it is useful to distinguish alternative indicators to impact and output metrics, namely *indicators of activity*. These suggest that it is possible to measure the effort that TTOs invest in engaging with non-academic users.

MEASURING TECHNOLOGY TRANSFER ACTIVITIES: ACTIVITY INDICATORS

Measures such as the level of patent registration only present a partial picture of the level of innovation. Other complementary indicators exist, or are being developed (by OECD and EUROSTAT), but these can pose problems in terms of availability, comparability and interpretation. One such indicator is the *technological balance of payments* (TBP), which measures international trade in technical knowledge and services – for example, on sales of patents, licences for patents, know-how, models and designs, trade-marks, and technical services. Such data is extremely valuable and sheds important light on a country's ability to sell its technological know-how, or conversely its dependence on importing foreign technology (Forfás, 2004a: 8). Understanding activities and processes is critical. Not all inventions or innovative activities result in a patent, and not all patents are exploited economically. Patents do not measure innovation *per se*, but rather the existence of knowledge that has the potential for innovation; these two things are very different.

Recently, research has called for a degree of flexibility to be introduced in respect of the measurement of university commercialisation and technology transfer activities (see Molas-Gollart *et al.*, 2002). Evidently, there is a need to heed the warning given by the OECD (2004): 'one should be cautious in the use of metrics and statistics, in that patents do not equal products and the number of licenses or licensing royalties do not represent the full extent and wide range of knowledge transfer that is actually transferred'.

The extent and impact of commercialisation activities cannot be based simply on a mechanistic application of measurements and associated funding formulas. Too much focus on patents as a basis for evaluating performance can be misleading, and may be inconsistent with the main missions of the universities or PROs. Rather than income and royalties, the focus of evaluation should be on the ability of TTOs to expeditiously and fairly negotiate and sign deals (Meyer, 2003: 15). More attention, therefore, should be paid to the processes that enable technology transfer. Molas-Gollart *et al.* (2002) provide a conceptual framework for analysing third-stream activities. They use broader conceptions for measuring commercialisation encompassing not only commercialisation in terms of patents, but commercialisation of university facilities. They also take a broad view of technology transfer including student placements and executive development. The validity of such an approach is the mutual inclusive and broad nature of measures rather then a narrow focus on outputs. The rationale for this focus provided by Molas-Gollart *et al.* (2002: 4): 'what is required is identification and measurement of the wide variety of processes through which universities engage society, moving well beyond a narrow focus on commercialisation activities that are relevant to only a subset of academic disciplines with clear industrial applications'.

Figure 5.6: Molas-Gollart *et al.*'s (2002) Framework for Analysing Activities

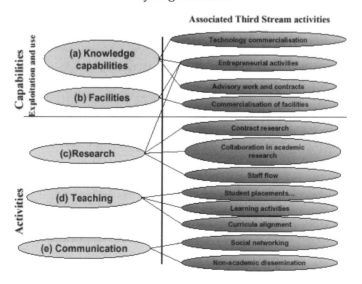

Evidently, measuring the impact of universities on society requires considerable effort and investment. Universities do not have the resources to conduct such wide-ranging studies on a comprehensive and regular basis. Instead, the approach suggested by Molas-Gollart *et al.* (2002) focuses on the fundamental elements of third-stream activities that universities can see for themselves. Focusing on the performance of activities facilitates in overcoming the problems associated with impact assessment measurements and serves as a feasible complimentary evaluation technique. This utility stems from the fact that activities can be measured and assessed, as opposed to outcomes or impact, which are complex and difficult to attribute causality to.

Measuring Technology Transfer: A Contextual Approach

The problem of using some of the common variables for technology transfer is that they are exogenous to direct influence by TTO staff. Output variables are contingent on a multitude of other variables – for example, quality of research being conducted, general economic conditions, and company policies. Such intervening variables, coupled with the complexity of the innovation and commercialisation process generally, as well as extensive time lags, serve to make evaluation of activities extremely difficult.

Research by Siegel *et al.* (2003) adjusted estimates of relative efficiency to reflect environmental and institutional factors that can influence the rate of technological diffusion at a given university. This was done by means of the following formula:

Technology transfer = f (INT, ENV, Org)

Where: **Internal** is a vector of inputs (average annual license agreements or revenue; average annual invention disclosures (a proxy for the set of available technologies); average annual TTO employees; average annual external legal expenditures).

Environment represents environmental factors, such as legal fees incurred to protect the universities' intellectual property and the legislative environment.

Organisational is a vector of organisational variables – for instance, the presence of a medical school and the public status of the university may be important institutional factors, as may differing stakeholder expectations.

While relatively simplistic, appreciation of contextual variables may explain the limited diffusion of best practice and serve as a mechanism for appreciating local contingencies, in terms of explaining variances in the relative performance of TTOs. Further, the importance of measures will differ from each stakeholder perspective. Based on a survey of 113 universities in the US, **Table 5.6** indicates that licenses and royalties were identified as outputs of technology transfer by a substantial majority of TTO directors and university administrators. Managers and entrepreneurs also frequently mentioned licenses, but stressed informal aspects of technology transfer also (Siegel *et al.*, 2003: 43). It is important, therefore, for TTOs to measure a number of variables and to communicate and emphasise the correct metrics to various stakeholders.

Table 5.6: Outputs of University-Industry Technology Transfer, as Identified by Interviewees

Outputs	Type of Interviewee		
	Managers / entrepreneurs	TTO Directors / Administrators	University scientists
Licenses	75.0	86.7	25.0
Royalties	30.0	66.7	15.0
Patents	10.0	46.7	20.0
Sponsored research agreements	5.0	46.7	0.0
Start-up companies	5.0	33.3	10.0
Invention disclosures	5.0	26.7	5.0
Students	25.0	26.7	15.0
Informal transfer of know-how	70.0	20.0	20.0
Product development	40.0	6.7	35.0
Economic development	35.0	20.0	0.0
Number of interviews	**20**	**15**	**20**

Based on a review of the literature in this area, Bozeman (2000) provides a model of the various contingencies that may influence technology transfer. The model considers a number of determinants of effectiveness, including various characteristics of the technology, the transfer agent and the technology recipient. The most important point of the model is that technology transfer effectiveness can have several meanings, including market impacts, political impacts, impacts on personnel involved and impacts on resources available for other purposes and other scientific and technical objectives research. Bozeman (2000) argues that, because of these complex relationships, it is very difficult to calculate the returns, economic and social, to academic research.

The model is useful, in that it draws awareness to the multitude of factors that may influence technology transfer. By appreciating such contingencies, and the various barriers and stimulants to technology transfer indicated in this chapter, TTO/ILO staff gain a richer, more in-depth understanding of the factors that shape the parameters of their activities. This understanding can be useful in communicating and managing the expectations of stakeholders.

Figure 5.7: The Contingency Effectiveness Model of Technology Transfer

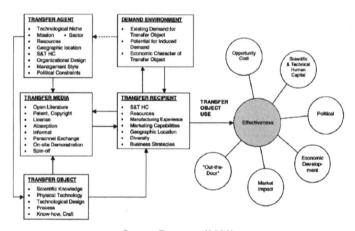

Source: Bozeman (2000).

IMPLICATIONS FOR HEIS

The key issues highlighted in this chapter that need to be considered by HEIs include:

♦ Stimulants for technology transfer and research commercialisation, which include **macro-level** factors, such as:

◊ The level of business investment in R&D in the region

◊ HEI proximity to industrial clusters

◊ Support of development agencies/research councils

◊ Institutional and legislative context

◊ Government support

◊ Features of the patenting process.

♦ **Micro-level** factors, which include:

◊ Research tradition

◊ Strategy, mission and clear objectives
◊ Top level leadership and commitment
◊ Quality of TTO
◊ Age and experience of ILO
◊ Scope of commercialisation initiatives
◊ Documented policies
◊ Educational offerings
◊ Knowledge of research and relations with faculty
◊ Networking and informal relations
◊ Trust and common expectations
◊ Inventor involvement
◊ Motivation, incentives for researchers and university culture
◊ Realistic expectations.

The three types of barriers to commercialisation of research are institutional barriers, operational barriers and cultural barriers:

♦ Institutional barriers:
  ◊ TTO organisational structure and supporting infrastructure
  ◊ Resource constraints and expertise deficiencies
  ◊ Poor internal relations
  ◊ Lack of clarity over ownership of IP
  ◊ Perceived conflicts of interests
  ◊ Over-estimating the value of IP
  ◊ Patenting of research tools may discourage beneficial research
  ◊ Partners may lack understanding or trust
  ◊ Lack of support for SME technology transfer
  ◊ Complicated technology transfer policies.
♦ Operational barriers:
  ◊ Lack of space
  ◊ Lack of investment in R&D by companies
  ◊ Lack of funding for prototype development
  ◊ 'Research treadmill'
  ◊ Availability of researchers to pursue research fellowship
  ◊ Constraints for researchers
  ◊ Contract research and salaries
  ◊ Lack of full-time research positions
  ◊ Confidence in commercial merits of the research.

- ◆ Cultural barriers:
  - ◊ Department and faculty attitudes towards commercialisation
  - ◊ Lack of awareness
  - ◊ Perceived non-challenging nature of commercialisation
  - ◊ Lack of soft skills.

Evaluating Technology Transfer

- ◆ The true impact of technology transfer is difficult to measure quantitatively (Fassin, 2000: 40). Caution must be exercised against measurements that focus on revenue-generating activities of TTOs.
- ◆ Introduction of indicators cannot be seen to be proposed merely for their role as performance metric but rather must be promoted to staff for the way in which the indicator will facilitate university administrators in the support and management of activities. Moreover, metrics should attempt to represent the full extent and wide range of knowledge transfer that is actually transferred.
- ◆ Consideration should be given to the following factors when developing metrics to measure technology transfer activities:
  - ◊ Need for 'local' metrics
  - ◊ Organic nature of university-industry interactions
  - ◊ Differences across disciplines
  - ◊ Resistance/lack of enthusiasm for evaluative metrics
  - ◊ Interdependence between university activities
  - ◊ Indirect impacts
  - ◊ Serendipity
  - ◊ Over-emphasis on efficiency *vs* effectiveness.
- ◆ Quantitative measures for TTOs could include:
  - ◊ Research expenditure
  - ◊ License option agreements
  - ◊ Research funding
  - ◊ License income
  - ◊ Legal fees, expenditures and reimbursements
  - ◊ Patent-related activity
  - ◊ Start-up companies
  - ◊ Licensed technologies and post-licensing activity.
- ◆ There needs to be a complimentary set of evaluation mechanisms that evaluates technology transfer activity *and* impact.

6

VARIETIES OF EXCELLENCE: BEST PRACTICE CASES IN TECHNOLOGY TRANSFER[19]

INTRODUCTION

In this chapter, we present varieties of excellence from technology transfer initiatives in three countries: the US (University of California), Israel (Hebrew University of Jerusalem) and the case of regional co-operation in Finland. These best practice cases serve to illustrate the diverse institutional arrangements for managing technology transfer activities and the various methods used to foster successful technology transfer. Each case has been chosen for specific strengths that may help illuminate the topic – for example, the University of California is one of the most successful entities worldwide at technology transfer and commercialisation, with its vast infrastructure and network consisting of 11 campuses. The Israeli examples highlight what can be achieved when all actors in the innovation process combine for the mutual good, while the Finnish case emphasises that a centralised policy with regard to a national system of innovation can help engender regional technology transfer activities through either the regional co-operation model or the technology transfer company model.

Each case in this chapter is structured as follows: we begin with a brief introduction, before moving on to the background and institutional context specific to the country being examined. Then we examine in detail the commercialisation and technology transfer process and highlight the intricacies and particulars in each of the cases, thus allowing us to draw an idea of how the cited 'best practice' examples function.

---

[19] This chapter was written by Will Geoghegan, a Research Fellow, Centre of Innovation & Structural Change, National University of Ireland, Galway, email: will.geoghegan@nuigalway.ie.

TTOs & Research Commercialisation in the US

Policy-makers and researchers continue to be attracted to the success of leading US universities in performing technology transfer activities (Molas-Gollart *et al.*, 2002). In order to review the successful technology transfer operations at the University of California, it is necessary to consider the institutional context, which differentiates the technology transfer activities in the US from other countries.

Institutional Context

The transfer of research results from government-funded R&D at federal and academic institutions to the private sector has grown significantly in the past two decades in the US and, today, it represents an increasingly important part of the overall industrial commercialisation of technologies. Federal Reserve Chairman, Alan Greenspan, recently noted (BHEF, 2001) that 'in a global environment in which prospects for economic growth now depend importantly on a country's capacity to develop and apply new technologies, our universities are envied around the world. The payoffs – in terms of the flow of expertise, new products and start-up companies, for example – have been impressive'. This is demonstrated by the fact that the rate of start-up formation is between three and four times higher in North America in comparison to most other OECD countries (OECD, 2002).

Since the passing of the Bayh-Dole Act, universities have become more involved in transferring technologies, with the ability to retain title and to license their inventions providing a healthy incentive. Paralleling the development of the university infrastructure for protection and licensing of intellectual property, there has been increased interest from commercial entities in forming partnerships with universities to exploit research. Industry in the US recognises that university research laboratories use specialised facilities and staff that cannot be obtained readily elsewhere. Universities also exhibit a dependency on industry in order to cope with revenue pressures, as a result of fund cuts by the federal government.

Allied to the Bayh-Dole Act in facilitating commercialisation and patent policy in the US, a number of factors have contributed similarly to the development of technology transfer policy on both a macro- and a micro-level. MIT and Stanford University pioneered technology transfer efforts at universities in the 1940s and a number of federally-funded R&D labs were created to meet the needs of World War II. In the 1950s, the government started to support small businesses with growth capital. The ambitious space program during the 1960s supplied more federal funding for R&D and required co-operation between government and industry to succeed. In the early 1980s, federal technology transfer became widely regarded as a means of addressing concerns about US industrial strength and competitiveness. The 'threat' from Germany and Japan contributed to the reforms that make up the foundations of today's technology transfer environment: a uniform patent policy, the access to venture capital and increased participation of small businesses in R&D and technology transfer. The 1990s saw the development of comprehensive educational programs on entrepreneurship and innovation at many universities, as well as the success of technology transfer resulting from the intense focus on biotechnology and medical R&D. Today, technology transfer remains high on the science and technology policy agenda in the US.

Research Commercialisation at US Universities

Since 1980s, the US has established a strong national technology-licensing infrastructure, which includes more than 200 TTOs (an increase of 800% since 1980). Some of the most successful institutions, which are now world-renowned for their technology-transferring skills include Stanford University, MIT, Columbia University and the University of California System (AUTM, 2002).

The main tenets of the US system of technology transfer include:

♦ TTOs are typically located within the office of the Dean of Research or have a direct reporting relationship to the Dean of Research. Some State universities in the US manage their research administration activity in external organisations solely because state laws do not permit universities to engage in such activity.

♦ Generally, activities within TTOs are governed by a set of policies, including guidelines for patenting, licensing, equity ownership, copyright, consulting and contracts with industry. Policies have been developed to protect the rights of researchers and to preserve core academic values, as well as to protect the university from conflicts of commitment and conflicts of interest.

♦ Mandatory assignment of inventions is an important policy at US universities. Employees at the University of California, for example, are required to sign a patent agreement in which they agree to disclose all potentially patentable inventions and to assign all rights of their inventions to the university.

♦ Public universities, including the University of California, have a public service mission to ensure that research results are made available for public benefit. Industry involvement is encouraged and it is becoming increasingly recognised that such contacts actually benefit academic research activities.

♦ Royalty arrangements vary but, typically, the loyalty share to the researcher is around 30% of royalties above $100,000. In some cases, early rewards are used to encourage the researcher to bring the technology along the commercial route. Typical early rewards involve a share to the researcher of 60% for the first $50,000 and 40% to 50% for the second $50,000.

University management of research commercialisation is still developing in the USA, despite their international lead. Most university policies are still somewhat *ad hoc*, with administrators and policy-makers learning by doing. However, certain norms and trends are emerging, such as the mixed skill set among TTO personnel, and the consolidation of these offices by segregating the operational elements.

Licensing & Patent Activities in the US [20]

US universities and colleges collected about $830 million in the financial year 2001 from royalties and other payments coming from licenses on inventions, according to a survey of 143 institutions by the Association of University Technology Managers (AUTM, 2003). This was between 2% and 3% of the total R&D effort conducted at universities and colleges that year. About 15% of all revenues went to a single institution, Columbia University, which reported royalties of almost $130 million. MIT ranked second, with $74 million dollars in revenues, half of which came from cashed-in equity from two of the institute's companies – the Internet-company *Akamai* and the biotechnology firm *Pracecis Pharmaceuticals*. It is important to note that this 'ranking' is highly dependent on a few 'blockbuster' licenses, which generate most of the revenues.

Revenue from technology transfer was not a large portion of overall R&D income at universities and colleges. Only 10 institutions reported royalties in excess of $20 million and 17 institutions over $10 million. Of approximately 23,000 active licenses reported in FY 2001, only 131 generated more than $1 million in income that year. The overall revenue from licenses was lower in FY 2001 in comparison to 2000, but technology transfer activities, in terms of invention disclosures (11,529 in 2001) appeared to be on the rise. **Table 6.2** shows the number of US patents issued to specific institutions during 2002. To put these figures in perspective, the IBM Corporation received 3,288 patents in 2002, more than any other private sector organisation, for the 10th consecutive year (USPTO, 2003). In a recent study of over 100 US research universities, the average number of invention disclosures during the 1990s was 67, with a mean annual license income per disclosure of $43,000 (Lach & Shankerman, 2003).

---

[20] Largely drawn from Karlsson (2004).

Table 6.1: Licensing Revenues Collected by Universities & Colleges in FY 2001

Rank	University	Licensing revenues ($m)	% of total R&D effort
1	Columbia University	130	37
2	MIT	74	17
3	University of California system	67	2
4	Florida State University	62	54
5	Stanford University	39	8

Source: AUTM (2003) and NSF (2003).

Table 6.2: Top Five US Universities Receiving the Most Patents (Preliminary Numbers) for Innovations during 2002

Rank	University	No. of patents
1	University of California system	431
2	MIT	135
3	California Institute of Technology	109
4	Stanford University	104
5	University of Texas	93

Source: USPTO (2003).

A key feature of the technology licensing offices at the large American universities is their size. For example, the TLO at MIT employs 17 technology-licensing specialists. This scale allows MIT to employ staff with a wide range of experience and to have a staff base qualified to work proficiently with a variety of technologies.

Start-Up Companies & Business Incubators

Many universities have set up special programmes for creating new businesses from their innovations. These programmes generally bring together service providers, funding sources and entrepreneurs. Stanford University and MIT serve as examples of the most successful universities in developing these kinds of practices.

Most universities offer at least a basic level of assistance. This includes help in preparing business plans, liaising with potential investors and providing an advisory service. A number of universities also offer incubator space or have arrangements with science or enterprise development parks in the locality. For example,

Northwestern University has established the Northwestern University Evanston Research Park, with the city of Evanston. A key driver or motivating factor for universities to support spin-offs is the desire to ensure that the region around the university directly benefits from the research undertaken at the university.

The University of Oregon puts spin-offs through a three-stage incubation process. The first stage provides the academic-entrepreneur with pre-incubation space, consisting of four offices where the faculty member can locate themselves to plan their business. At the next stage, the spin-off project moves to an innovation centre incubator, where R&D can be carried out or small-scale production can be done. The final stage sees the spin-off project move to a research park space. Projects move to this stage when they are too large for the innovation centre and less assistance and support is required. Throughout this process, there is a lot of help and expertise afforded to the academic entrepreneur. A business consultant is employed to train spin-off promoters and provide advice. The programme also involves start-up companies linking with student programmes. For example, teams of MBA students can be assigned to assist a start-up to develop its business plan and carry out market analysis. Support is also provided by the law school on issues such as determination of product liability.

The following section will provide a more in-depth review into the TTOs at the University of California (UC), which is one of the most successful universities at the practice of technology transfer. As a major public university, the UC is often considered a model that sets standards for other universities, nationally and internationally.

University of California: Office of Technology Transfer

> *We seek cooperative relationships with industry not simply to generate royalty revenue and stimulate economic growth, but to create relationships with industry that will help faculty in pursuing their own research and in training graduate students.*

Richard Atkinson, president of the University of California

Mechanisms for supporting the commercial exploitation of any university-researched patents were put in place at the University of California (UC) in 1943, and assignment by faculty of their inventions to the university was determined on a case-by-case basis. As a result, the UC, which comprises 10 campuses, is the leading university system in the US in terms of number of patents and in the number of successfully-commercialised inventions. In FY 2002 (the year ending June 30, 2002), a record number of 973 inventions were disclosed from all campuses. Inventions in life sciences, including medicine and biotechnology, accounted for over 70% of new inventions.

University of California: Mission & Objectives

One significant aspect of the University of California's public service mission is to ensure that the results of its research are made available for public use and benefit. This transfer of technology is accomplished through various methods, including education of students, publishing results of research and ensuring that inventions are developed into useful products in the commercial marketplace for public use.

In pursuit of this latter objective, the University of California has maintained an active patenting and patent licensing program for over 40 years. The main objectives of the University patent programme are:

♦ Disseminating new and useful knowledge resulting from university research through the use of the patent system.

♦ Licensing patents to industry, in order to promote the development of inventions toward practical application for use by the general public.

♦ Providing income for use in supporting further research and education, with a share of the income going to the inventor; and to ensure that patent-related obligations to the sponsors of research are met.

University of California: Evolution in Technology Transfer

While the UC has a long history for commercial exploitation, it was in 1980, as part of a broader expansion in UC patenting and licensing activities, that the university staffed its office (the Patent, Trademark and Copyright Office – PTCO) with experts in patent law and licensing. The Board of Patents was abolished in 1985, and new policies allowing for the sharing of patent licensing revenues were adopted by the Office of the President and the campus' Chancellors in 1986. Staff employment in the PTCO grew from four in 1978 to 43 in 1991, when the PTCO was renamed the Office of Technology Transfer (OTT). In 1990, UC Berkeley and UCLA established independent patenting and licensing offices, relying on the system-wide OTT selectively for expertise in patent and licensing regulations. By 1997, four UC campuses (UC San Diego and UC San Francisco, in addition to Berkeley and UCLA) had established independent licensing offices.

The University of California has continuously developed and formalised its technology transfer activities, raising awareness among its staff of the commercialisation options available to them. This is best reflected by the increase in the number of invention disclosures by researchers as documented in **Figure 6.1** below. It is also interesting to note that the increase in disclosures predates the passage of the Bayh–Dole Act. Furthermore, as the University of California sought to develop the 'top 5' inventions, the focus shifted from agricultural inventions (including plant varieties and agricultural machinery) to biomedical inventions.

Figure 6.1: Disclosures at the University of California, 1976-1989

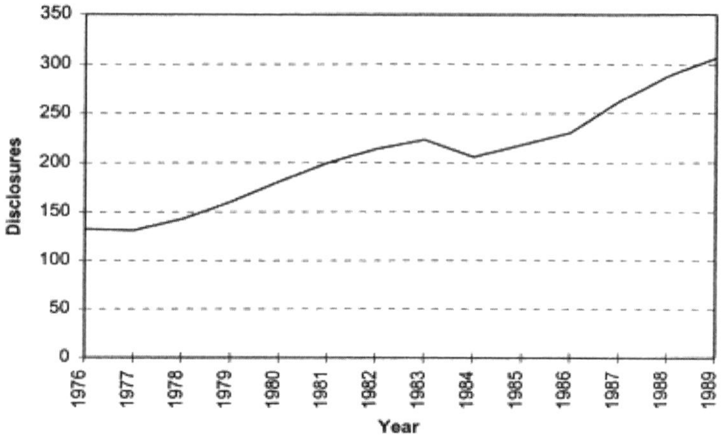

Source: Mowery et al. *(2001).*

TTO Structure

The UC system represents 11 campuses, including several that would stand alone as prominent research universities – for example, UC Berkeley, UC San Francisco, UC Davis and UCLA. Nonetheless, a single central Office of Technology Transfer exists at the University of California's Office of the President, which co-ordinates technology transfer activities for the entire system, in cooperation with autonomous licensing offices at the larger campuses. The OTT oversees system-wide efforts to encourage the commercialisation of results. It does this under a 'distributed' model. Through this model, the OTT co-ordinates and supports technology transfer activities across all 11 campuses.

At a system-wide level, the OTT provides policy development and guidance, legal support, system-wide information management, legislative reviews, accounting and a variety of other co-ordinating services in support of the system-wide programme. The OTT also provides a comprehensive management of a substantial invention portfolio and offers a wide range of infrastructure services to emerging and well-established campus and laboratory transfer offices. Until

1994, the OTT held the central responsibility for technology transfer activities. However, in 1994, the Ad Hoc Technology Transfer Advisory Committee recommended that the technology transfer should be faculty-centred, inventor-centred and campus or laboratory-centred. Six campuses established their own licensing offices and three campuses have chosen to rely solely on the licensing services of the OTT, and each is assigned an OTT technology transfer liaison. Each campus and laboratory develop and shape technology licensing programs that are suitable to their particular needs, under a memorandum of understanding negotiated through the UC Office of the President. General oversight of the UC technology transfer program is provided by the Technology Transfers Advisory Committee (TTAC). This standing committee advises the UC President on technology transfer policy and guides the direction of the overall programme.

Each campus' Office of Technology Licensing (OTL) reports to the Vice Chancellor of Research. For example, the OTL was established at Berkeley in 1990 and reports to the Vice-Chancellor for Research. The Berkeley website (http: //otl.berkeley.edu/about/about.php) notes: 'We work with campus inventors to facilitate transfer of technologies created at UC Berkeley into the commercial sector for public use'. The scope of these activities include:

♦ Evaluating the commercial potential of new technologies.
♦ Determining patentability.
♦ Prosecuting patents.
♦ Registering copyrights.
♦ Marketing and licensing patents, tangible material, and software.
♦ Negotiating license agreements.
♦ Receiving and distributing royalties and other income to the inventors, UC Berkeley Campus and its Departments.

An important objective for UC is to promote the wide dissemination of new ideas to the general public. The OTL's staff of patent, copyright and licensing professionals is available to the entire campus community for commercial evaluation of new technologies and inventions for patenting, registration of software copyrights, marketing and licensing of these intellectual properties.

Management of Inventions

A number of services are run centrally to assist industry in identifying the services available to them on the different campuses. The OTT acts as a one-stop shop for potential industry clients wishing to access any of the 11 sites under its remit. The OTT also uses the Internet and various databases to further university research and transfer technology relationships with industry. Companies throughout the world use the system-wide OTT home page to access the comprehensive database with abstracts of federally-supported university research in progress. Industry uses the database, which is updated monthly, to locate university faculties for research collaborations and sponsored project support. Searchable information on research units, centres, and institutes are also included on this site. In addition, this site allows companies to search for technology licensing opportunities on all UC campuses and laboratories.

At the campus level, each campus retains an office on-site and overall activities and policies are determined centrally. In recent years, each campus has assumed greater responsibility for technology transfer activities. The campus offices work with the OTT to develop a unique programme to meet the goals and requirements of their own campus. Each campus is committed to a different level of technology transfer activity, in agreement with the Office of the President and the OTT. During 1999, the initial five-year pilot period for campus-based technology transfer at San Diego and Irvine was completed. This pilot period was deemed successful and has been extended indefinitely. Los Angeles operates a co-operative arrangement with the OTT, where the OTT will assume greater responsibility for managing patents, licensing and accounting activities on behalf of the campus. Technology transfer specialists at the OTT, and on each of the campuses, are required to have a strong technical and subject-area background, experience in negotiating with industry and knowledge of intellectual property issues.

Revenue & Income

The total UC licensing revenue reached $100 million in FY 2002. UCSF alone, generated more than one-third ($34 million) of that amount. Of the total licensing revenue, $88 million (up 21% from FY 2001) came from royalties and fees derived from 980 technologies. The top 5 commercialised inventions earned royalties exceeding $40 million.

Principles Framework

The principles policy was developed in 1999 to address issues regarding rights to future research results in university agreements with external parties. Following increasing interaction with industry during the last decade, agreements have become more complex and the number of different types of agreements has increased. However, there are a set of common values at the university that should be preserved and they must be agreed upon, articulated and communicated.

The principles offer a basic framework that enables UC to maintain consistency in managing research results across the campuses, while providing greater flexibility in local administration of agreements. In addition, the principles were designed to provide university negotiators with a basis to support positions taken during often-challenging contract negotiations.

Table 6.3: Principles Framework at the University of California

1. Open dissemination of research results and information.
2. Commitment to students.
3. Public benefit.
4. Informed participation.
5. Legal integrity and consistency.
6. Commitments concerning future research results.
7. Fair consideration for university research results.
8. Objective decision-making.

These principles apply to all university agreements with external parties, including contracts and grants, which affect rights to research results, according to the policy. The University of California was one of the first universities to develop a principles policy and it has generated a lot of interest from other universities.

Supporting Technology Transfer Activities at UC

The UC San Francisco (UCSF) is one of the world's largest health-sciences institutions, a prominent actor in biomedical research, patient care and higher education. With 18,000 employees and an annual budget of about $1.9 billion, UCSF is San Francisco's second-largest employer. UCSF scientists have been responsible for launching over 70 California biotechnology companies since the 1970s, including industry giants *Genentech* and *Chiron*. Starting in 1969, UCSF institutionalised the practice of collaborative research, by grouping researchers with a common interest instead of creating specialised departments. This strategy established a cross-disciplinary 'hothouse' – an innovation machine (*Business Week*, 2003). One result is that UCSF has, by far, the largest patent and license agreement portfolio of all UC campuses. The entrepreneurial environment at UCSF includes the Entrepreneur Discussion Group (EDG), which provides an informal environment for getting feedback on early-stage ideas, and the UCSF Centre for BioEntrepreneurship (CBE), which was established in 2001 to introduce students and faculty to the business skills needed for success in the life science industry. In addition, the UCSF Innovation

Accelerator (IA) supports the growing entrepreneurial spirit by establishing a network of Bay Area scientists and business professionals interested in life science entrepreneurship and helps entrepreneurs to develop their innovative ideas into viable businesses.

A good example of the collaborative ventures happening in UCSF is a new type of consortium called *PharmaSTART*, led by *SRI International*, which was announced in August 2003. Apart from UCSF, three other leading California research organisations were founding members: Stanford University, UC San Diego and the UCSF campus of the California Institute for Quantitative Biomedical Research (QB3). PharmaSTART will help accelerate the translation of breakthrough new drugs from discovery into clinical use. The consortium will address this innovation gap by offering drug development and consultation services, create new collaborations, start drug development initiatives and find new funding sources (see http://www.sri.com).

The UC San Diego (UCSD) also has a strong position in biomedical sciences. Among all medical schools, UCSD School of Medicine ranks first in the US in federal research funding per faculty member, and six departments are in the top 10 in National Institutes of Health (NIH) funding in their areas. The health sciences attract nearly 42% of UCSD's total research funding. In the 1980s, the university started to develop close university-industry linkages and has become a significant player in entrepreneurial development in the San Diego area. The university has maintained a regional orientation toward licensing and many licensees are in the San Diego/Orange County/Los Angeles area. UCSD has spun off 150 San Diego companies, including 63 in the biomedical industry. To help speed up the commercialisation process, its Technology Transfer & Intellectual Property Services (TTIPS) has established a License & Entrepreneur Assistance Program (LEAP) that helps small companies in the community, as well as entrepreneurial faculty researchers, to access and commercialise UCSD technologies.

The CONNECT Program

One of the America's most famous technology transfer initiatives is the *CONNECT* program. Established in 1985, it is primarily an economic development organisation focused on regional high-technology entrepreneurship. According to the organisation itself: 'UCSD CONNECT is widely regarded as the most successful regional program in the US linking high-technology and life science entrepreneurs with the resources they need, such as technology, funding, markets, management, partners and support services' (http://www.connect.org/). The program has assisted more than 800 technology companies since it was established, and is entirely self-supporting, receiving income from members, events and grants but not university funding. Nearly 1,000 companies support the program in one way or another. This model has been replicated in other cities and countries, including Sweden and the UK.

CONNECT has a wide range of activities on its agenda, including networking activities, events, courses and assisting entrepreneurs. One example is a program called *Springboard*, which has a focus on early-stage entrepreneurs. Participants receive coaching in business plan development by experienced technology executives. When ready, participants make a presentation to a feedback panel of investors, lawyers and industry representatives. Springboard has helped over 150 companies to raise more than $200 million in capital. CONNECT also hosts an annual Biotechnology Corporate Partnership Forum. The forum provides an opportunity for San Diego biotech companies to introduce their technologies to pharmaceutical companies, venture capitalists   and   larger   biomedical   firms   (see   http: //www.connect.org/programs/springboard).

TECHNOLOGY TRANSFER & COMMERCIALISATION
IN ISRAEL

Over the past 20 years, Israel has transformed itself from a traditional agrarian economy to a country with high technology industry, currently growing at an exponential rate. Various sources of evidence indicate that this is indeed the case, including:

♦ In recent years, Israeli scientists have had the highest number of patents issued *per capita*. In addition, more scientific articles are published *per capita* by Israeli scientists than by any other country.

♦ Many leading MNC technology-based companies have their research facilities in Israel.

♦ Israel has one of the world's highest percentages of engineers and scientists in its labour force.

♦ In 2000, Israel was second only to the US in annual investment in high-technology start-ups.

♦ Israel's knowledge-based companies are concentrated in a few industries: telecommunications, software and biotechnology.

♦ According to IMD's *World Competitive Yearbook* (2000), out of 47 countries, Israel ranks 11th in 'science and technology' and 8th in 'people' or Human Resources.

Universities, supported by a focused innovation system, are considered to have had a significant input to this transformation. A substantial proportion of basic research in Israel is done by universities. While only 10% of civilian R&D resources in Israel are allocated to universities, a disproportionately large share of the basic research is done there. This has led Meseri & Maital (2001: 15) to comment that 'a reasonable hypothesis, then, is that the heart of Israel's technology transfer developments is the process in which knowledge is transferred from where it originates – Israeli universities – to where it is commercialised, the private sector'.

An important component of the universities' mandate is to develop research undertaken by their researchers for the industrial development of Israel. This best practice case will examine broadly the infrastructure that enables such technology transfer in Israeli universities, before focusing on one particular example of best practice, the Hebrew University of Jerusalem.

Institutional Context

The impetus for a focus on R&D can be seen to stem from when the French government imposed an embargo on the export of technology to Israel, which led to a strategic decision to reduce dependency on foreign technology by improving the Israeli R&D capacity. This policy was further facilitated by the influx of highly-educated immigrants, mainly from the Soviet Union, into Israel and a reduced spending in military-based industries, which pushed people into civilian industries.

The law for the encouragement of Industrial Research & Development (1984) governs all public research in Israel. The law seeks to encourage companies to invest in R&D, with the government sharing in the risks involved in such projects. Under this law, when government-assisted projects result in a product or process with commercial potential, the developers are obliged to pay royalties to the funding agency. The law also requires that these royalties be used to support further R&D.

Co-operation between academia and industry is encouraged by the Ministry of Trade & Industry, through its Chief Scientist, by programmes such as *Magnet*, which aims at achieving a more efficient allocation of limited financial and professional resources through co-operation between companies and academic research institutions. Furthermore, in addition to domestic research schemes, Israel also involves itself in research schemes with other countries – for example, the USA-Israel Science Foundation, the Technology Commission and the Canadian-Israel Industrial Research & Development Foundation.

Intellectual property law in Israel advocates that all inventions made by a researcher, whether patentable or not, are owned by their employer, be it a university or a public research laboratory. All the

Israeli universities have developed strong research commercialisation and technology transfer units. These units are generally well-resourced and have strong links with other components of the Israeli innovation system, including industry. The operation of these technology transfer units are described in the following section.

Technology Transfer: Some Case Evidence

Meseri & Maital (2001) undertook a survey of the technology transfer organisations that operate in six of Israel's universities, consisting of:

♦ B.G. Negev (Ben-Gurion University of Negev).

♦ Bar Ilan Research & Development Company (Bar-Ilan University).

♦ Dimotech (Technicon-Israel Institute of Technology).

♦ RAMOT (Tel Aviv University).

♦ Yeda (Weizmann Institute).

♦ YISSUM (Hebrew University).

Research findings indicate that the key criteria used by TTOs to accept or reject invention disclosures proposed by faculty members were those shown in **Table 6.4**.

Table 6.4: Criteria Used to Assess Invention Disclosures by TTOs in Israel

Criteria	Average Score
Market need	5
Market size	4.3
Existence of patent	4.3
Success chances for R&D	4.2
Degree of maturity of the idea	4.1

Source: Meseri & Maital (2001).

Meseri & Maital (2001) found that Israel's TTOs apply the same criteria as used by their counterparts in the private sector to assess technology transfer projects or invention disclosures – a focus on market-based criteria rather than a perceived national need. They also questioned technology transfer operations in Israel as to the factors they perceived as being essential success factors for effective technology transfer; the results are shown in **Table 6.5**.

Interestingly, the perceived success of a technology transfer is noted as being strongly related to the quality and motivation of the research team. The main focus of the TTO's staff is concentrated on licensing, with only one of the six offices placing a stronger emphasis on new ventures/start-ups.

Table 6.5: Perceived Success Factors for Technology Transfer

- ◆ Real and agreed need for the project's innovation.
- ◆ Clearly-defined project goals.
- ◆ Demonstrated market demand and ability to penetrate the market.
- ◆ Adequate and stable financing.
- ◆ Degree of consensus.

Source: Meseri & Maital (2001).

Technology Transfer Companies

Of the seven universities in Israel, six have a technology transfer company, which promotes commercial exploitation of the institution's research, and provides consultancy services to industry. In most cases, the technology transfer companies have limited liability and are wholly-owned by the university. In other cases, the ownership of these companies is across more than one university. The mechanisms differ from university to university and the following section outlines the practices of each of these companies in terms of technology transfer.

Technion Research & Development Foundation Ltd

Technion Research & Development Foundation Ltd (TRDF) is a limited liability company established in 1952 to manage the research activity of the Technion, which itself serves as a centre for basic and applied research in the sciences and engineering and is considered one of the top institutions in the world for research in these disciplines. Technion has 40 research centres, a large number of which engage in high-technology research in engineering, science and medicine. Technion has a student body of over 13,500 and completely owns TRDF, which is responsible for the management and administration of sponsored research; industrial testing; external and continuing studies; and the commercialisation procedures for know-how, inventions and innovative technologies generated at Technion laboratories.

The commercialisation process at TRDF begins by ensuring that the legal rights of the owners of intellectual property are protected (this is usually achieved by filing patents). Before a patent application for a technology is submitted, the TRDF Commercialisation Committee reviews the technology. This Committee consists of representatives from the TRDF, Dimotech, Technion Entrepreneurial Incubator

Company (TEIC) and a representative from TRDF's research authority. The Commercialisation Committee decides on the most appropriate course of commercialisation (see **Figure 6.2**).

Figure 6.2: The Commercialisation Process at TRDF

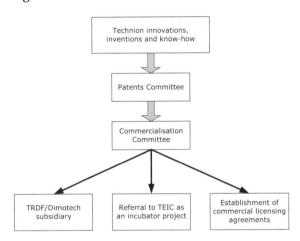

The Business Development & Financial Control Department co-ordinates all of the commercialisation activities described above, including the implementation of legal and other procedures. Its responsibility ranges from initial contact between the Technion scientist and the commercial entity, through designing the agreement, defining ownership of intellectual property, and figuring out how the intellectual property is going to be shared between the parties when commercialised. Activities include:

♦ Analysis of new inventions.

♦ Filing of patent applications; protecting and maintaining IP rights.

♦ Negotiation and approval of the IP and business aspects of agreements with industry and joint industrial-academic projects financed by companies and/or by the Chief Scientist's Office of the Ministry of Industry & Trade.

♦ Licensing of IP.

♦ Incorporation of spin-off companies based on Technion IP, both in TRDF and in TEIC.

♦ Participation on the board of directors of TRDF-affiliated companies.

Dimotech Ltd

Described by its managing director as a 'holding company for start-up industries at the Technion, specialising in commercialisation of R&D projects', Dimotech's set-up includes 22 high-tech companies in the fields of medicine, computers, electronics, biotechnology, agriculture, energy and economics. Dimotech was set up to enable projects through the R&D stage and to develop their commercial potential. Dimotech accomplishes this by developing alliances between companies, investors and/or strategic partners with viable R&D project groups at the Technion. It also commercialises and licenses patented innovation and nurtures the start-up companies that set out to develop, manufacture and market technologies developed at the Technion. Dimotech also has a role identifying scientific projects that are at a progressive stage and ready to embark on commercialisation.

Technion Entrepreneurial Incubator Company

The TEIC is also a wholly-owned subsidiary of the Technion and comprises 2,000 square meters of laboratories, offices and workspaces, providing potential academic entrepreneurs from the Technion with facilities to commercialise their research through spin-off companies. Projects supported by the TEIC typically receive incubator organisational benefits, including:

♦ Assistance in attaining seed and development funds from public and private sources.

♦ Assistance in preparing business plans and strategies for financing, manufacturing and marketing their inventions or know-how.

♦ Assistance in locating strategic partners and investors.

♦ Access to administrative support, offices and workspace.

Ownership in companies owned by TEIC is divided between the academic entrepreneurs, the workforce, TEIC and investors. To date, the number of enterprises established by TEIC is approaching 50.

RAMOT University Authority for Applied Research & Industrial Development Ltd

RAMOT University Authority for Applied Research & Industrial Development Limited is the technology arm of Tel-Aviv University (TAU), Israel's largest university with a student body of over 25,000 students in nine faculties and 106 departments.

RAMOT was founded in 1973 by TAU with the mandate to:

♦ Register, administer, and commercialise patents and inventions originating from the university.

- ◆ Explore and develop avenues for the application of research carried out in TAU towards solving specific industrial problems.
- ◆ Couple industry with academic research, innovation and expertise.
- ◆ Provide technical and managerial know-how to institutions and industry.

RAMOT grants companies licenses to the intellectual property developed at TAU. It also deals with the patent-filing decisions, marketing the technologies to prospective licensees, negotiating licenses with companies and monitoring licensing agreements. RAMOT generally takes a flexible approach when negotiating licensing agreements, tailoring the agreements to fit the specific research and business needs of a particular project. RAMOT holds title to the inventions and intellectual property derived from TAU research, although, as an incentive to researchers, a substantial share of the net income is allocated to the inventor to continue research in the same area. Companies that have included licensing agreements with RAMOT range from new start-ups to large global corporations.

RAMOT also serves as the contact point for contract research at TAU and actively participates in the MIT *Magnet* programme. TAU researchers participate through RAMOT in nine *Magnet* consortia over a number of fields. RAMOT serves as TAU's formal representative on the consortia and coordinates the participation of the researchers from TAU. In addition, RAMOT operates an on-line information database, where consortia members can access data, as well as post research results to other members and the general public.

RAMOT also serves as an active holding company that establishes spin-off companies and technology incubator firms. The underlying principle at TAU and RAMOT is to allow entrepreneurial researchers to maintain involvement in the commercialisation of their innovation, while continuing their academic work at the TAU. RAMOT also manages one of the 26 technological incubators in Israel at its incubator unit, RAD-RAMOT. This allows the spin-off enterprises to qualify for financial assistance from the Israeli MIT.

The Case of YISSUM (Hebrew University)

Since its inception 75 years ago, the Hebrew University has been strongly involved with the needs of society and has contributed widely to the enhancement of society's welfare. The university's mission has been, and continues to be, to achieve excellence in research and teaching, to contribute to society and play the role of cultural and academic. When it comes to the organisational development for the management of university-industry linkages, the university controls all contractual matters very closely and directly enforces its policy and regulations.

Organisational Structure

The Hebrew University established in 1964 a wholly-owned subsidiary, called the 'Research Development Company of the Hebrew University of Jerusalem' (YISSUM), to foster relations between the university's researchers and industries. YISSUM's aim is to create business arrangements with industry throughout the world to commercialise innovations of the Hebrew University. YISSUM currently holds 400 registered patents and has 150 projects in operation.

On its establishment, the university empowered YISSUM to act on its behalf with regard to any contractual matter with industry relating to the universities know-how, patents and research results. YISSUM acts as an independent and separate legal entity, providing services to the university in the area of university-industry management.

As with other universities, YISSUM is the exclusive owner of the university's intellectual property and is charged with the promotion and commercialisation of the university's applied research for public use and benefit. The contractual relations between the university and the company are based on university regulations in the area of university-industry interface.

YISSUM's Operational Model

YISSUM commercially develops selected applied research by evaluating the technologies, identifying potential markets, and negotiating university-industry cooperation. It seeks to do this in a manner that does not interfere with the normal academic activities of lecturing and basic research.

YISSUM has, or has had, marketing agreements with a variety of US and international companies, particularly in the pharmaceutical and chemical industries, including Ciba-Geigy, American Cyanamid, Sandoz, Schering, Merck, Pfizer and Johnson & Johnson. For marketing technologies, its main focus is on the US market, where it finds it easier to market and negotiate licenses. YISSUM accepts that, once a particular technology is adopted, other research organisations will move into this area, usually with significant manpower and financial resources. YISSUM estimates that there is usually a six-month window of opportunity for many innovations. Taking account of this, YISSUM encourages its researchers and research groups to bring forward a continuing steam of new ideas for development.

Other than publishing and disseminating abstracts through quarterly newsletters and press releases, YISSUM does not pursue joint ventures pro-actively. Instead, it relies on its academic researchers to bring forward initial contracts. This is sufficient to generate over 100 signed agreements per year, although many of these are relatively small. YISSUM also links with the Israel Trade Missions in different countries for identifying potential customers for its licenses.

Structure of YISSUM

The president of the university nominates the board of directors of YISSUM. Board members include businessmen and a handful of university employees. This Board then elects the managing director of the company.

YISSUM is organised around four main areas of operation:

♦ Business development.
♦ Marketing and database management.
♦ Patents and portfolio management.
♦ Contracts management.

Companies transfer to the university all funds received from industry for research conducted by a university researcher. It is YISSUM's responsibility to collect these funds, based on formal agreements.

YISSUM also charges the industrial partner 35%, in addition to funds agreed on, for continuous or complementary university research. This overhead is split as follows:

♦ 7.5% to YISSUM to cover its operational costs.
♦ 27.5% to the university.

The Interface

Defining the interface, its organisational structure, its tasks and charter is an important prerequisite for YISSUM's efficient functioning within the university. Building an interface as an independent entity enables it however, to become a source for internal change of attitudes. The interface and its successful ventures with industry have provided a positive spirit towards cooperation with industry, including researchers whose research topics are mainly basic. The interface is also involved in exhibitions demonstrating achievements in applied research, both to the internal university community and to industry, venture capital funds and investors. Such exhibitions create an atmosphere for greater co-operation between university and industry and, when made on campus, will also influence students, enhancing their comprehension of interaction between industry and university. This atmosphere of co-operation adds competition on campus and fosters more inventiveness and creativity in both applied and basic research. Experience from YISSUM suggests that an appropriately-balanced atmosphere, encouraging further relations between industry and university, leads to the creation of a dynamic environment to create and establish courses with industry or of industrial orientation.

RESEARCH COMMERCIALISATION IN FINLAND[21]

Before the collapse of the Soviet Union in 1991, approximately half of Finland's exports went to the USSR. This collapse, and the simultaneous worldwide recession in the early 1990s, had a major impact on Finland, with unemployment increasing dramatically. In 1994, to address this situation, the Finnish government moved to develop Finland as a globalised information society, while at the same time initiating a programme of restructuring and market deregulation. A key component of this programme has been the development of a national system of innovation. As a progression of this programme, science and technology are now at the top of the national agenda.

Research Funding in Finland

In the 1990s, the research system in Finland was overhauled and, as a result, funding for R&D rose to 3.1% of GDP in 1999, one of the highest in the OECD. Between 1995 and 1999, public funding rose by 40% to FIM 7.6 billion. The rapid expansion of the system is also shown by the increase in the number of R&D workers, from 42,000 in 1993 to 61,000 at the beginning of 2001. Coupled with the increase in numbers, there has also been a significant increase in the standard of education (see also **Chapter 1**, for a comparison with Ireland).

In 1998, Finnish universities spent about €590 million on research with about 45% of this investment coming from external sources. The contribution by external funding sources increased both quantitatively and proportionately in the 1990s, with the most important sources of being the Academy of Finland and TEKES (the National Technology Agency). The international contribution to research funding has also increased, primarily because of EU funding programmes.

Over the last 10 years, cooperation between the universities and the business sector has become much closer. Targeted research funding and various other mechanisms have advanced the dissemination of research results for use by business. Special emphasis has been placed in university funding on rapidly-developing growth areas that are also

---

[21]   Typically, discussion in relation to Finland centres on the importance of Nokia in explaining its success to date. See, for example, Kaukonen & Nieminen (1999) for a useful contribution to this debate.

important for the business world, such as information and communication technology and biotechnology.

The Finnish Innovation System

Finland has a well-established, comprehensive and highly-interactive national innovation system. The key objective of the system is to implement an approach that strengthens the competitiveness of basic industry, while also developing new high technology industries. Key organisations in the system include the Ministry of Trade & Industry, the Ministry of Education, Academy of Finland, National Technology Agency (TEKES), the Foundation for Finnish Inventions, SITRA and the universities.

Figure 6.3: The Finnish Innovation System

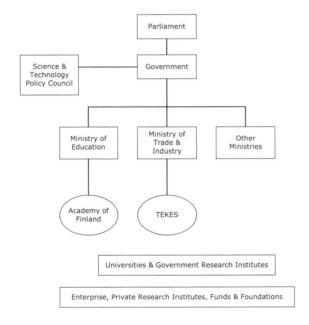

Source: European Commission (2003).

The Ministry of Trade & Industry has the primary responsibility for Finland's industrial and technology policy and for the creation of the preconditions for the development of Finnish industry and enterprise. The Academy of Finland is an expert organisation on research funding and operates within the administrative sector of the Finnish Ministry of Education. The objective of the Academy is to seek to enhance the quality and reputation of Finnish basic research through systematic evaluation and by influencing science policy. In 2001, the Academy allocated over €150 million for research funding, about 12% of total Finnish government research funding, to 3,000 researchers working on Academy-funded projects.

TEKES, the National Technology Agency of Finland, is the principal organisation responsible for technology policy in Finland. TEKES supports companies engaged in risk-bearing product development projects with grants and loans, and finances the projects of research institutes and universities in applied research. TEKES' technology programmes have proved to have a key role in starting co-operation between companies, universities, research institutes and other public organisations. The programmes concentrate on areas where R&D investment on a national level is important.

The Foundation for Finnish Inventions supports and helps private individuals and SMEs to develop and exploit inventions. It has an especially important role in supporting the early stages of the exploitation. The aim is that as many projects as possible are then transferred to others to finance further and can be commercialised.

SITRA, the Finnish National Fund for Research & Development, is an independent public foundation under the responsibility of the Finnish parliament. Research funded by SITRA focuses on future challenges facing Finland. The research is intended to build a store of know-how for decision-makers and national debate. It is applied, multidisciplinary and directed towards the future.

TTOs' Commercialisation of Research in Finnish Universities

Finland operates in a different legislative context to the US, in that the inventors own the patents. The first attempts of technology transfer from university to industry began in Finland in 1992, as part of a nationwide initiative. In 1999, after slow progress, the Finnish government started a project in the six main universities to increase the volume of deals. The main goal of the three-year project was to strengthen international licensing know-how in Finland.

There are two main models for commercialisation research in Finland. The first, and more common, model for research commercialisation and technology transfer, is for the university to work closely with other agencies in the region. The second model is where the university has its own technology transfer company. Case studies of both models are presented in **Table 6.6**.

Table 6.6: Cases in Regional Co-operation & Company Models

I. Regional Co-operation model	II. Technology Transfer Company model
The Technical University of Helsinki has close links with the Oteniemi Science Park (OSP). OSP employs specialists to regularly survey the activities in R&D in the universities and research institutions in the region. Projects and researchers who show interest and potential in the commercialisation of research qualify for entry into a first-stage incubator. The incubator evaluates the feasibility of the business idea and its possible further development, notably by offering training and consultancy.	Abo Akademi University Foundation and the University of Turku take an approach much similar to that of the previously-mentioned US and Israeli models.
The core to the OSP is the Spinno Business Development Centre, the development and educational unit of OSP. It is supported by TEKES, and Colminatum Ltd Oy (Helsinki Region of Expertise). Spinno BDC evaluates the start-up projects for their application and potential for intellectual property protection. It also provides enterprise training to the project promoter and facilitates cooperative interactions with other agencies. This is done on a nil-cost basis for the first 20 hours. About one-third of the projects are then offered premises and incubator status.	Aboatech Ltd was established in 1933 and is jointly owned by Abo Akademi University Foundation, the University of Turku Foundation and SITRA. Aboatech seeks to identify, evaluate and protect new commercially-promising innovations originating from Abo Akademi University, the University of Turku and Turku University Hospital. The main focus of Aboatech is on technology transfer and the commercial exploitation of high-technology innovations through licensing and selling of products both in Finland and abroad.
Spinno BDC then provides them with financial advice and assist them to procure financial support, including government grants from TEKES. If the enterprise exhibits growth, Spinno BDC can offer additional support in terms of advice. Spinno also operates a mentor network of about 100 retired and experienced businesspeople.	Aboatech has an additional role of promoting Turku's universities. It liaises with Spinno, TEKES, the Foundation for Finnish Inventions and other agencies on behalf of the two universities. It takes a strong marketing approach to its technologies and many technologies are licensed overseas.

The Tampere University of Technology has also established two companies – *Hermia* and *Finn-Medi* – to enable the commercialisation of university intellectual property. They both have well-established incubator programs, and are now trying to establish a licensing business. Hermia was established in 1992 and has 18 people. Over the last 10 years, more than 200 start-ups have been created there. Finn-Medi handles life science inventions from the University of Tampere, Tampere University of Technology and Tampere University Hospital, and works in cooperation with Hermia. They were both among the first university technology licensing organisations in Finland.

As part of their developing and training process, both Hermia and Finn-Medi recognise the importance of a complete understanding of the technology transfer process. To facilitate this, they formed a collaborative effort with Stanford University, whereby staff from Hermia and Medi each spent 12 weeks working at the Stanford Office of Technology Licensing. This learning process involved working alongside staff at the Stanford OTL and observing how the office operates, how technologies are evaluated and marketed and how to negotiate licenses. This relationship has built up over time and, subsequently, a number of the staff from Stanford visited Finland to gain an appreciation of the technology transfer process in a Finnish context and to provide them with a contact in Europe.

Promoting University Industry Interaction in Finland

Two case studies prepared by Klofsten & Jones-Evans (2000), although somewhat dated, are interesting because of the fact that they focus on university-industry interaction rather than on research commercialisation directly (see **Table 6.7**). Evidently, the nature of broader university and industry interactions will ultimately impact on the success or otherwise of research commercialisation activities.

Table 6.7: Cases in University-Industry Interaction in Finland

The Turku Technology Centre: An Interface between Academia & Industry	The Adjunct Professor Model in Tampere
The Turku Technology Centre, in south-western Finland, is a major centre of new technology and is occupied by more than 100 enterprises and 30 university institutes, departments and laboratories. The majority of the enterprises in the centre are involved in information technology, electronics, biotechnology and materials science. In addition, the University of Turku, Abo Akademi University, the Turku School of Economics & Business Administration, as well as other research and technical organisations, have units in this centre. The centre has resulted in improved co-operation between the universities and industry. There is an increased volume of joint work and a number of new enterprises and new projects have been spun off. Universities and companies even have established joint laboratories within the centre, including the Centre for Biotechnology, the Turku Centre for Computer Science and the Materials Research Centre. Technology transfer is one of the important activities of the Turku Technology Centre. This is organised by Aboatech Ltd (see **Table 6.6**), which serves researchers, universities and companies by creating contacts and networks. Using national and international co-operation networks, Aboatech helps researchers assess the possibilities of commercialising their ideas and find partners to exploit them.	The Tampere University of Technology does not have a centralised university-industry liaison function and, instead, retains direct connections with industry through the academic staff. One of the more recent forms of industrial collaboration has involved the employment of part-time professors (or adjunct professors). These professors share about 20% of their time with the University. The selected industrial professors mainly lecture to students, thus giving them a close view of the input into the R&D process. Academic entrepreneurship and technology transfer are not the objective of the initiative, rather the key objectives are to enhance relations between industry and the university, provide industry perspectives in educational courses and provide industry commentary on R&D underway within the university. The university currently has three chairs of this nature through Nokia, UPM Kymmene and IBM.

Key Lessons from Varieties of Excellence Cases

A number of observations can be made from the review of TTO arrangements for the commercialisation of research in the US, Israel and Finland. It must be stressed that these efforts should be referred to as 'varieties of excellence', in the acknowledgement that institutionally-specific circumstances will mean that only certain elements or a mix of elements will be particularly relevant to any one TTO. This section, therefore, will take a cautious approach; not generalising from exceptions but rather proposing key success factors and commonalities among the cases discussed.

At each TTO reviewed, the management of the interface for university-industry relations is clearly derived from the university's policy. Despite the various organisational forms that the TTOs have taken, there exist a number of commonalities in the cited cases, which can be seen to lead to a perceived set of 'best practice' criteria, including:

♦ **Resource allocations:** The offices are well-resourced with specialist and expert personnel. The TTOs reviewed recognise the importance of having staff with the appropriate skill mix; many of the specialists have industry experience or marketing backgrounds.

♦ **Transparent policies:** Offices are run by clear and transparent polices. This was most evident at YISSUM where, despite its separate legal status, activities were strictly governed by university policy and regulations.

♦ **Pro-active approach:** Offices are pro-active in auditing the research at their respective institutions and aim to identify research with commercialisation or technology transfer personnel. Close personal contact and building relationships with researchers may help facilitate in developing trust, raising awareness and ensuring that the TTO is up to date with research developments.

♦ **Taking control of technology transfer:** The TTOs are involved in controlling most elements of the technology transfer process and are increasingly developing their abilities as marketing offices for the institution's technologies and are pro-active in identifying potential recipients of these licenses. In Israel and Finland, this focus is particularly directed towards international markets.

♦ **Leveraging networks:** None of the offices operate as stand-alone entities but rather leverage networks, or associations (AUTM in the US, for example) for knowledge and expertise. At the University of California, general oversight of the technology transfer program and advice on larger decisions is provided by the Technology Transfers Advisory Committee (TTAC). In Finland, TTOs are very much embedded in the national innovation system, while even the

University of California, which is generally regarded as the one of the most well-resourced institutions in the US, is involved in the CONNECT network and consortiums such PharmaSTART (in co-operation with Stanford University, UC San Diego and the UCSF campus of the California Institute for Quantitative Biomedical Research).

♦ **Focus on licensing and patenting:** It is evident that research commercialisation is heavily focused on technology transfer *via* licensing and patents rather than enterprise development. This reflects the fact that there is a receptive market for licenses (both Finland and Israel have a strong focus on the US market for their licensing activities). In the US, this facet has become wide-scale, due to moves by most universities to exploit technology transfer opportunities afforded to them by changes in legislation – for example, the Bayh-Dole Act.

♦ **Visible and active supports:** Research commercialisation is also seen to be stronger where there is active and visible support for such activities. This is deemed to be central for developing a culture that is supportive of technology transfer and commercialisation activities. As with every important initiative, support and commitment from the top management of the university is essential to the success of the TTO. Towards this end, the board of the university should approve an internal policy for intellectual property rights and for technology transfer and set up the necessary infrastructure to support technology transfer activities.

♦ **Rewards:** At both YISSUM and UC, there are clear incentives and motivations for researchers to disclose their inventions. Revenues generated by the university by licensing its inventions are typically shared among the inventors, the inventors' laboratories/ departments and the university's general fund. The potential for invention disclosure may also be enhanced by the education and awareness programs evident at the cited institutions. Israeli academics, for example, have a positive disposition towards research commercialisation and working with industry in general. This has been facilitated through working in academic-industry consortia for a variety of research projects (particularly under the *Magnet* programme). This disposition is also assisted by cultural traits of working towards the good of the state.

♦ **Managing strategically:** Strategic prioritising can be seen as an influential criterion for the potential success of a university's commercialisation prospects. The cited examples suggest that a key focus for TTOs is to decide how to allocate development efforts, since not all disclosures offer an equal promise of success (Allan, 2001). In Finland, special emphasis in university funding has been on rapidly-developing growth areas that are also important for the business world, such as information and communication technology and biotechnology sectors.

♦ **Quantitative measures mask some sustainable performances:** A major stumbling block for a successful commercialisation programme is that, at best, quantitative measurement of a TTO's activities is problematic. It is evident that there is little correlation between budgets, number of patents, or other quantitative measures and the degree of success a TTO will have in terms of generated income. Columbia University, for example, has often appeared at, or near, the top of AUTM's survey in terms of licensing income generated. Yet, it was granted an average of only 34 patents per year from 1994 to 1998.

To conclude, this chapter by no means attempts to point to a universal optimal configuration or structure for a technology transfer but one may see from the cited examples that certain common threads exist between the successful initiatives. Through deep analysis of these and other success stories, one may garnish a fuller appreciation into what may constitute a successful technology transfer or commercialisation programme.

7

A STRATEGIC APPROACH TO TECHNOLOGY TRANSFER

The reunion of industry and science through new technology effects a radically different, much more intimate combination of forces than anything that has been obtained before.
Burns & Stalker (1961: *xxi*)

INTRODUCTION

Historically, there has been a failure to capitalise on the true value of HEI-business collaboration and thus realise the potential force outlined some time ago by Burns & Stalker (1961). Throughout this book, attention has been drawn to the changing roles and demands being placed on HEIs. Successful HEIs in the 21st century will have to excel in teaching and research, as well as in developing complementary third-stream activities to exploit research for the public good. The ultimate criterion of success will be the quality of the service that a HEI provides. Given their multiple roles as information brokers, science marketers and catalysts for academic entrepreneurship, TTOs provide a crucial mechanism to allow HEIs respond to this challenge (Fassin, 2000: 41).

While international evidence from best practice presented provides a basis for understanding the requirements for success, individual HEIs will have to find their own solutions to these issues, depending on their size, disciplinary base and the economy of their region (Hall, 2004; Shattock, 2001). The basis of such policy development, therefore, should acknowledge the 'variety of excellence' that exists and allow each HEI to choose its own distinctive strategy (Molas-Gollart *et al.*, 2002). As the OECD notes 'there is no one-size-fits-all approach to technology transfer' (OECD 2003: 13). Therefore, it is better to speak of a configuration that draws on elements of best practice elsewhere, acknowledging that policy choice and implementation will have to be adapted to suit local circumstances and contingencies. In developing such an approach, HEI leaders will need to give due consideration to the development of a strategic approach to technology transfer at their institution. **Figure 7.1** presents the key issues in doing so.

Figure 7.1: Developing a Strategic Approach to Technology Transfer

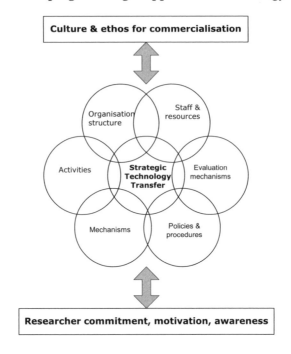

The central issue in **Figure 7.1** is that HEIs must take a more pro-active strategic approach to commercialisation and technology transfer activities rather than merely playing a facilitator/information broker role. The circles surrounding the core of Strategic Technology Transfer indicate the interrelated mechanisms that will support and facilitate this strategic approach.[22] Crucially, this diagram acknowledges that such changes cannot take place without due consideration for the motivation and awareness of researchers and a supporting ethos for commercialisation in terms of the broader context of the HEI. In exploring these issues further, we will draw on the criteria that were identified as crucial to the successful development of technology transfer initiatives, while attempting to address the potential barriers to commercialisation that were highlighted previously (see **Chapter 5**). Ultimately, improving technology transfer and commercialisation are

---

[22] It must be acknowledged that some of the issues suggested will be connected to the types of pressures that emanate from Government agencies and funding bodies. The authors, however, recommend a pro-active approach to implementing such policies rather than having them imposed. The rationale is that successful transfer stems from experience and older policies. Hence, the earlier policies are put in place, the more benefit in terms of long-term outcomes.

keys to an innovative economy, leading to wealth generation and job creation (Large *et al.*, 2000).

An Issue of Balance

Crucially, while commercialisation of property owned by HEIs is an important component of third-stream activities, it is only one among the many other functions that link HEIs with society (Molas-Gollart *et al.*, 2002). In this respect, it is worth remembering:

♦ The naturally occurring, organic nature of HEI-industry interactions.

♦ The fact that third-stream and second-stream activities are not independent.

♦ The considerable difference between HEIs across disciplines in the ways that research influences society.

Appreciation and acknowledgement of these issues can facilitate in preventing any potential backlash by researchers or staff, who may feel commercialisation activities are being imposed on them (OECD, 2003). Essentially, the issue is one of balance. The emphasis should be to balance the objectives of managing intellectual property rights, developing new revenue sources, and accommodating faculty interests, while simultaneously maintaining norms related to the conduct of academic research and the dissemination of research findings (Feldman *et al.*, 2002; OECD, 2003: 104).

Frequently, research has shown that the priority for the TTO is getting the deal done rather than attempting revenue maximisation or waiting for a better deal. The key issues, therefore, are presented on the basis that they are not an attempt to alter the underlying basis of academia but rather to advise on the creation of a suitable environment with the appropriate conditions to stimulate those researchers who wish to commercialise their work. For those who voluntarily choose the commercialisation route, this should be as effective and as attractive as possible. The multiple and mutual benefits of ensuring this option is available were highlighted in **Chapter 1**. The following sections provide an in-depth analysis of the key issues central to the development of a strategic approach to technology transfer.

KEY ISSUE 1: STRATEGIC MANAGEMENT OF
TECHNOLOGY TRANSFER

The emergence and continual development of TTOs at most HEIs in an *ad hoc* and unstructured manner is evidence of their acuteness and vital necessity. The downfall of such a 'passive', rather than strategic, approach to the exploitation of IP, however, tends to be lower levels of patenting and licensing, coupled with poor infrastructural support and low levels of awareness and commitment to commercialisation activities (OECD, 2003: 42). HEIs should therefore develop a more pro-active, strategic approach to technology transfer, including the development of a mission statement, a strategic plan and the identification of strategic priorities.

Key Success Factor 1:
Development of a strategic plan, mission statement and strategic priorities.

Developing Strategic Plans

As HEIs become more involved in commercial activities of one kind or another, they will have to develop clearer ideas of their mission and their objectives, as well as clear policies to facilitate implementation of these (for example, for dealing with potential conflicts of interest). As a result of HEIs' far-reaching role, activities are often uncoordinated and subject to an unnecessarily burdensome system of accountability and regulation. Consequently, there is requirement for the parameters of future development of technology transfer initiatives to be documented in a strategic plan. This plan should define the goals and mission for technology transfer and, in so doing, should provide a sense of direction and leadership, as well as a clear rationale for action. As one of our interviewees noted: 'there needs to be a *total* ... institutional strategy for industry partnerships within institutions, not the piece-meal approach that is currently common on a department and college basis'.

In other words, HEIs must provide the organisational support to attract, negotiate, and carry out collaborations with industry. This needs to be assessed and documented in order for various parts of administrative structures to work together in an effective manner to reach common goals. It is critical to link objectives clearly into the HEI's strategic and academic plans, and demonstrate how activities will facilitate in achieving aspects of the HEI's own mission.

While TTOs are more complex to manage than businesses, they need to be business-like in the way they manage their affairs (Lambert,

2003). Evidence from the Hebrew University of Jerusalem and the University of California indicates that one of the critical success factors in the management of HEI-industry linkages was an institutional strategy for the development of such relationships, laid down in a strategic plan. Strategic management of relationships implies, not only the formalisation of policy priorities, but also top-level management support (Martin, 2000). The commitment of the institution's leadership is essential to the long-term success of technology transfer initiatives. A strategic plan serves to galvanise, and provides evidence of leadership and commitment. Such evidence can be particularly useful in contexts where HEI technology transfer is controversial. Once a plan has been devised, the institution's leadership will also play a key role in communicating the objectives and emphasising the importance and relevance importance of industry collaborations and technology transfer (Meyer, 2003).[23] Aside from communication on web-sites and speeches, as well celebrating success stories, successful HEI change programmes have involved an annual letter to the research faculty and staff, written by the Chancellor and President, highlighting the importance and appropriateness of involvement in technology transfer (Meyer, 2003).

Crucially, the process of developing the key objectives and mission statements must be an all-inclusive one. Acceptance and agreement of objectives by multiple stakeholders is of crucial importance, and also serves to facilitate the process of implementation. Reviewing best practice in technology transfer, Allan (2001) noted that development or reform of policies should involve inputs from funding agencies, scientists and industrial partners, as well as other actors in the technology transfer chain. Stakeholder involvement not only will create buy-in to the newly-established objectives, but also will ensure that there is a consensus in terms of expectations of the costs, risks and time-lag associated with commercialisation activities.

Objectives arising from the planning process should be clear and, while they must take historical perspectives and political realities that bear upon the particular institution into consideration, they should be as simple and straight-forward as the institution's circumstances permit. Understandable objectives in the form of a strategic document can ensure effective management and provide the TTO with legitimacy when dealing with internal and external stakeholders. A small number of important objectives (no more than seven to 10) also makes communication and dissemination of policy easier.

---

[23] Allan (2001) and Scott *et al.* (2001) note that one mechanism for such emphasis is that the language of technology transfer is an inherent part of public statements, publications and speeches made by, and on behalf of, the leadership of a HEI.

Table 7.1: Elements of the Strategic Plan for Technology
Transfer at Georgetown University

- New organizational structure.
- License intellectual property database.
- Commercial evaluation of inventions.
- Market technologies and license/abandon (18 months).
- Analysis and marketing of existing portfolio.
- New boiler-plate agreements.
- Develop start-up opportunities.
- Outreach to faculty and Washington region.
- Performance-based management.

Developing Mission Statements

The development of a strategic plan, however, can only take place in an effort to meet the aims and scope of activities as laid out as part of a mission statement, which, quite simply, should communicate what the TTO is and what it does. The statement should avoid any direct reference to the commercialisation of research for fund-raising or revenue-maximisation purposes as the key role of the TTO. The statement should serve to ease the fears of those who are critical of commercialisation, emphasising that technology transfer is a contribution towards fulfilling the overall mission of the HEI. Of the mission statements reviewed in **Chapter 3**, most stressed 'the exploitation of research for the public good or for public use'. Harvard's mission statement is a useful template:

> To bring HEI-generated intellectual property into public use as rapidly as possible, while protecting academic freedoms and generating a financial return to the HEI, inventors and their departments.

This statement neatly captures the dual responsibility of TTOs: first, as 'guardian of HEI intellectual property' and, second, its role in disseminating this intellectual property for the public good. Research by Siegel *et al.* (2003) concluded that mission statements were indicative of leadership and an entrepreneurial climate that was complementary to a HEI generating more licenses.

Overall, a 'strategic' focus on technology transfer should become manifest in the other interrelated mechanisms that support and facilitate this approach, including the identification of strategic priorities as elaborated upon below.

Key Issue 2: Technology Transfer Activities

Resource constraints acknowledged, successful TTOs have been noted for conducting strategic prioritisation in order to encourage and facilitate research. Allan's (2001) review of best practice suggests that Technology Transfer Officers should try and shed responsibility for everything except the key processes involved in technology transfer. In the literature this has been referred to as focusing on core competencies – that is, activities or areas where the institution excels (or has the ability to excel in).

Key Success Factor 2:
Focus more on action, as opposed to facilitation.

Identifying Technology Platforms

While the desire to manage all research activities with the same level of effort and commitment will always be to the fore, the reality is that most institutions identify areas of competitive strength in research. The salience of research in HEI portfolios has led to strengthening of strategic research planning within institutions. These efforts have been guided, in no small way, by the level and type of funding being allocated through government agencies. These are observations that carry particular weight in an Irish context, where the establishment of the Higher Education Authority's Programme for Research in Third Level Institutions and the creation of Science Foundation Ireland have encouraged HEIs to develop research bases, promoted research excellence and encouraged more high-risk fundamental research. Most institutions do not have the critical mass or expertise, or even resources, to conduct in-depth research in all areas of science and technology. There is a need for focus and for expenditure to be prioritised within an overall coherent framework that promotes national development objectives (ESG, 2004; ICT Ireland, 2004; Lambert, 2003). In this way, HEIs can build up clusters of knowledge in particular areas that lend themselves well to commercialisation, should researchers wish to follow this route.

TTO's Role

The TTO should have a central role in identifying and encouraging the development of such strategic research competencies, or what have been termed 'technology platforms'. The TTO should also be in a position to communicate elements of this focus and, in turn, to direct businesses making contact with the HEIs to these niches of expertise. Such technology platforms also provide a useful opportunity for collaborative research contracts, so that the funding and development of such technology platforms does not remain the sole responsibility of the HEI. It is often the case that technologies developed in such platforms are clearly associated with the research requirements of enterprise. This serves to enhance the potential for technology spillovers and increased networking between enterprises and the HEI.

Building Strategic Technology Platforms

Strategic technology platforms can form the basis for viable and sustainable technology transfer initiatives and can contribute substantially to ensuring the continuing relevance of research investment. Recent recommendations by the Enterprise Strategy Group and the Working Group Report & Recommendations *Commercialisation of R&D in the ICT Sector* (2004)[24] highlight the benefits of developing such platforms as being:

♦ Connecting different businesses together in networks or clusters of common interest.

♦ Articulating enterprise needs to the research and education communities, thereby defining specific applied research projects.

♦ Prioritising longer-term research needs.

Such activity should be encouraged and nurtured by government agencies, and funds should be directed to institutions attempting to pursue this goal. The promotion of such clusters and competencies should serve as a magnet to attract other enterprises that are interested in collaborating in a particular research or technology area. The potential benefits of strategic technology platforms are best captured by the Enterprise Strategy Group: 'the process of identifying strategic platforms fosters strong interaction and knowledge transfer between the different players in the innovation system' (ESG, 2004: 67).

Frontier Research

The Working Group Report & Recommendations *Commercialisation of R&D in the ICT Sector* (ICT Ireland, 2004) also acknowledged that the development of cluster-based research agendas will serve to target funding to strategic areas where Ireland can develop internationally recognised applied research competencies. Such recognition is becoming an inherent part of HEIs' broader strategic objectives, which provide clear guidance for TTOs. The strategic plan at NUI Galway, for example, stresses that the university will continue to review and prioritise research areas, while also stressing the need for effective

---

24 ICT Ireland (2004).

strategic planning of research activity (Strategic Plan, National University of Ireland, Galway 2003-2008: 29). Research is expensive in time, equipment and salaries and, in a highly competitive environment, it is natural that HEIs need to build up reserves that can be invested into research areas that match the new research economy. The rationale is that, because of blurring lines between 'basic' and 'applied' research, attention focus should be on 'Pasteur's Quadrant', or what some have termed 'frontier research'. Frontier research, because it is at the forefront of creating new knowledge, is an intrinsically risky endeavour that involves the pursuit of questions without regard for established disciplinary boundaries or national borders (HLEG, 2005: 11).

Figure 7.2: Pasteur's Quadrant

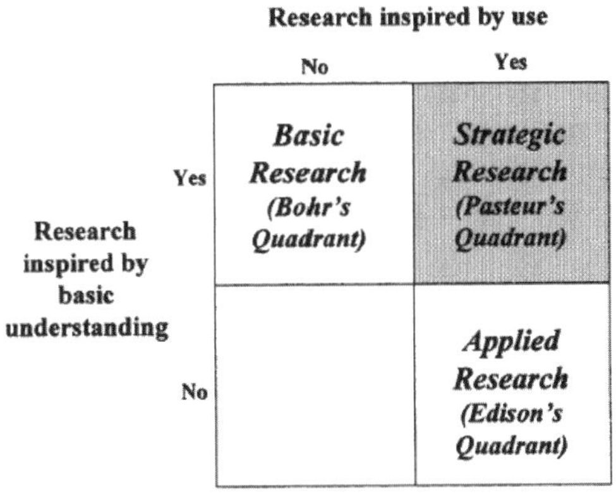

Research inspired by use

		No	Yes
Research inspired by basic understanding	Yes	Basic Research (Bohr's Quadrant)	Strategic Research (Pasteur's Quadrant)
	No		Applied Research (Edison's Quadrant)

Source: Stokes (1997).

A focus on frontier research implies a need to develop a strategic view of research activities. Shattock (2001) notes that this rationale has often underpinned the decision that HEIs have made to designate a vice-rector or pro-vice chancellor to take the lead, both in overseeing research offices and other organisational units, and in trying to develop a coherent institutional research strategy. The requirements for a strategic approach to encourage research activities is accentuated in the context of small economies, such as Ireland, which do not have the resources and expertise to compete in all emerging fields and, therefore, must be selective and specific around the areas in which it chooses to focus and invest. The ESG (2004: xvi) notes that 'developing an appropriate focus can be assisted by the identification of strategic technology platforms/areas of technology (for example, biometrics) that draw on basic fields of knowledge (such as mathematics, physics, and computing) for application to a wide range of products'.

Monitoring Research Activity

Closely-linked to the development of strategic technology platforms is the requirement to monitor actively the research activities of the HEI's researchers. While, traditionally, the academic has taken the lead in approaching the TTO for guidance, best practice suggests that the TTO should be active in monitoring the research being conducted in an institution. Although, technically, a step removed from the actual research, the TTO should have familiarity with the expertise of, and good contacts within, the HEI faculty and should track potential industrial research opportunities. Mechanisms that may facilitate this monitoring include the maintenance of databases of active research in the HEI, as well as areas of academic specialisms. These databases will be useful when it comes to negotiating and setting up collaborative research contracts or materials transfer agreements. Furthermore, such databases maybe useful internally, in terms of linking researchers who maybe looking at particular areas of common interest or with related benefits. Active monitoring is particularly warranted as HEIs extend the nature and extent of their research activities.

Developing & Sustaining an Active Network

A central mechanism in providing the opportunity for, and facilitating, technology transfer includes the development and sustaining of informal and formal contacts and networks. TTOs should leverage alumni networks to build closer relationships with graduates working in the business community. The informality of these contacts should not be judged as indicative of less effective contact mechanisms, but rather, as the OECD (2003) highlighted, informal contacts often form the initial underpinning to more advanced agreements. Linkoping University's SMIL centre believes that informal networking through informal talks and meetings has made a significant contribution to disseminating information regarding research commercialisation and enterprise development and has positively affected the culture and ethos at Linkoping University on these issues.

Develop a Presence in the Region

To maximise the potential of developing a presence in its region, a TTO should allocate time for staff to attend conferences and presentations and to mediate with government agencies and the local chambers of commerce to keep abreast of recent developments. Many TTOs retain databases of useful contacts and their areas of interest.

TTOs need to make the most out of networking opportunities, training programmes, and other membership benefits.

Horizontal Networking

Networking should not just occur with commercial entities but also horizontally between HEIs. Communication and discussion with other HEIs, and other TTOs in particular, should facilitate diffusion of best practice, as well as ensuring that staff are up-to-date with developments in various fields of research. This horizontal collaboration may also allow smaller HEIs to learn from larger, best practice institutions. Extensive networking may also lead on to the sharing of knowledge (for example, knowledge of the patenting process), or the sharing of resources in order to deal with common problems (for example, difficulties in interpreting legislation or facilitation in rectifying expertise deficiencies). Such a body of integrated knowledge and expertise may also prove more attractive to businesses, particularly multinationals. The frequency and extent of such interaction maybe formalised, if there is a sufficient critical mass and willingness to meet on a more structured and regular basis. Indeed, strategies of alliances and networking have become a key factor behind the success of HEIs (Lundvall, 2002). Therefore, there should be a maximum opportunity for networking, both formally and informally, between TTOs (particularly, their specialists in enterprise development and technology licensing).

In the US, the Association of University Technology Managers serves as one such mechanism. The AUTM organises and provides:

♦ Expertise in legislative and patenting issues.

♦ Regional meetings and programmes of study, providing intensive and cost-effective benefits for new and experienced members of the technology transfer community.

♦ A forum for debate and exchange of ideas.

♦ A journal and newsletter, documenting developments and trends.

♦ An annual licensing survey, which provides comparative data to enable benchmarking.

Another example of such a body in the US is the Licensing Executives Society (LES), which is a professional organisation that deals with all aspects of technology licensing. Interestingly, membership is from both corporate and HEI ranks. LES provides excellent training and continuing education programmes, in addition to opportunities to participate in issues relating to marketing, ethics, and international aspects of technology transfer. While the scale of activities in the US makes such bodies feasible, it is suggested that even the most basic form of interaction at a regional/national/supranational level would prove valuable in a European context. The major barrier here is the potential unwillingness of HEIs to co-operate. Clear communication of the benefits and efficiencies, as well as guidelines for operation, would reduce fears. Thus, the stereotype of the ivory tower has to be diluted somewhat, not merely because of increased interaction with industry, but also as a result of increased horizontal interaction between HEIs. TTOs need to be more strategic in their approach, moving from random individual contacts and one-person focus points, to making such interaction an inherent part of their activities and policies (Brennan & Shah, 2003).

Focus on Core Activities & Developing Core Competencies in these Areas

Research from best practice institutions stresses that, as TTOs develop, they need to become more strategic in their focus and to develop expertise in specific areas in which they wish to excel (Brint, 2005). In order to reduce the likelihood of being over-burdened and, thus, of missing commercialisation opportunities, there is a necessity to shed non-core activities so that resources and attention can be focused on previously-identified strategic priorities. Fassin (2001) emphasises the need for TTOs to focus on their core competency – *making deals*. Where possible, TTOs should minimise academic obligations and shed responsibilities for everything except technology transfer (MTAs, contracts, etc.). In the process of developing a strategic plan, activities should be prioritised in context of resource constraints and the previously-identified technology platforms and research priorities.

Research by Jones-Evans (1999) has highlighted the more sophisticated approach of Swedish TTOs. While TTOs still serve as one-stop-shops, in practice, this means that certain requests are referred to other offices/centres within the HEI, which have taken over and developed expertise in these area – for example, career placement, student placement or dealing with SMEs. HEIs and research institutions cannot be everything to everyone and must differentiate themselves and use their TTO to exploit the research that may prove most valuable.

Boundary-spanning: Communicating to Stakeholders & Managing Expectations

Best practice research of TTOs by Allan (2001) suggests that boundary-spanning activities by TTO staff are crucial in maintaining positive relationships with academics and can serve to reduce informational and cultural barriers to research commercialisation. Boundary-spanning refers to the specific type of marketing relevant to Industrial Liaison Officers.[25] Boundary-spanning capabilities and awareness ensure that TTOs act as an effective bridge between commercial entities and researchers. Research by Siegel *et al.* (2003) found that boundary-spanning on the part of the TTO is crucial and involves adept communication with stakeholder groups in order to forge alliances between researchers and firms.

Communication capabilities are an inherent part of boundary-spanning capabilities. Clear communication is critical for TTOs, given their interface role and requirements to serve multiple stakeholder demands and requirements. Communication from the boundary-spanning activities reduce the probability of conflict between different stakeholders, as they facilitate in forming clear expectations and an understanding of each other's role (Beesley, 2003). Boundary-spanning activities, therefore, prove an effective tool for managing communication and, consequently, can be crucial in obtaining buy-in at all levels of the HEI. Furthermore, effective boundary-spanning and relationship management enables trust to be built up between partners.

Development of Databases

One mechanism to facilitate and focus boundary-spanning activities is the use of software or a database to track and manage inventions, patents, agreements and contacts.

Duke University provides an excellent example of an institution that uses a common technology transfer database, in which all invention disclosures, patenting efforts and licensing activities and research agreements are tracked. Some HEIs – for example, John Hopkins – have adopted a software database (*FileMaker Pro*) to track all commercialisation activity for a technology, including information pertaining to ownership, assessment, marketing, licensing, royalty income, and legal fees.

Research by Sanchez & Tejedor (1995) has indicated the crucial developmental role of a database network, in that it allows staff to know about technology services and research activities available from every HEI department. Such moves towards planned organisation can reduce the chance element involved in meeting with contacts and commercialising research. In order to leverage potential contacts fully, some organisations keep databases of graduates employed by local companies. On a more macro scale, CORDIS, the EU Commission's Community Research & Development Information Service has created a technology marketplace, where businesses and researchers can

---

[25] 'Boundary-spanning' captures the importance of the networks and contacts of TTOs, as, in this context, advertising or technology brokers are rarely used.

search through and display research findings from EU-funded contracts that are available for licensing (Forfás, 2004a: 36).

Improved Industry Interface for Research

In the US, research organisations and HEIs make extensive use of the Internet to improve the interface between industry and research organisations (for example, University of California). Internet resources can be used to directly link research organisations with companies. Activities such as stakeholder mapping (in which stakeholders in each organization are identified), decision-making analysis (in which the decision-making processes of each organisation is made explicit), and expectation mapping (in which the expected roles and contributions of each partner are analysed) may also facilitate in managing the HEI interface, allowing for expectations and objectives to be understood better.

Outreach & Education

One of the key boundary-spanning activities TTOs should conduct is outreach and education programmes. Lessons from best practice institutions indicate that TTO activities should include facilitating periodic meetings, educational seminars, and training sessions for inventors and staff. Subjects can include business news, innovation theory, policy and organizational changes, and information resources. Educational activities are outlined in **Table 7.2**.

Table 7.2: TTO Outreach & Educational Activities

- Encouraging scientists to develop business skills through enterprise workshops and courses.
- Providing role models of success stories, and generating positive messages about the entrepreneurial culture.
- Stimulating interaction between industry and research through sabbaticals in industry for researchers, and by bringing industry people into HEIs as guest lecturers.
- Building awareness of what is needed to translate a research outcome into the marketplace.
- Putting in place appropriate industry-focused boards for R&D groups.

The goal of such activities is to develop attitudes and understanding of commercialisation. These efforts may be furthered by 'town hall' meetings/sessions involving all stakeholders. As noted by Meyer (2003: 14): 'to ensure success of the technology transfer office, its personnel should make outreach presentations on an ongoing basis and provide educational materials to teach the process of technology transfer to the faculty and student researchers'. Research has shown that such educational dissemination is best made to small groups in an informal interactive setting, such as a departmental or college faculty meeting. Presentation content can include the following:

◆ The technology transfer process.

◆ Best practice.

◆ Role of the Technology Transfer Office.

◆ Information on intellectual property rights.

◆ Names of faculty role models and examples of HEI inventions that have been licensed and successfully commercialised by industry.

Industry Visits

Other communication mechanisms include brochures containing information on the HEI's research strengths and organising visits by corporate customers to HEI research labs, as a way of highlighting HEI capabilities and facilitating personal contact between HEI and industry representatives.

TTOs, therefore, must spend a portion of their time engaging in boundary-spanning activities to encourage faculty members to disclose inventions. Often, the activities and purpose of staff of the TTO are not known, even within the HEI community. Contact and educational initiatives facilitate in creating an image for the TTO, which stimulates an awareness of the services provided and may create interest among researchers. Furthermore, technology transfer is a difficult and lengthy process, requiring patience and persistence. Education and outreach programmes can also serve as mechanisms to convey these truths to academics who have unrealistic expectations. Ultimately, the best marketing and communication tool is a good success story. Similarly, the best internal marketing for the TTO is a happy professor (Fassin, 2000: 36).

KEY ISSUE 3: ORGANISATIONAL STRUCTURE

Reflecting the *ad hoc* and rapid nature of the development of TTOs, very little attention has been directed at the organisational structure that may best facilitate technology transfer. Many existing TTOs have grown into their current structure, rather than taking a strategic stance to dictate their structure. Structural options, reporting levels and levels of autonomy are heavily shaped by institutional norms and path dependencies. The right mechanisms may depend upon the scale and substantive focus of the collaboration.

Key Success Factor 3:
There is no one best way to structure TTO activities.
Structure will be heavily shaped by institutional factors
and strategic focus.

The structure of TTOs at HEIs are complex and path-dependent (Bercovitz *et al.*, 2001; OECD, 2003). Historical appreciation of this is crucial; particularly, in estimating potential cultural resistance should structural alterations be attempted. The appropriateness of one institutional arrangement or another depends on the context within which the PRO operates: 'its status as a private or public institution, the amount of government funding it receives; the size of its research portfolio and fields of specialisation; its geographical proximity to forms and insertion in innovation networks; and its funding capacity' (OECD, 2002).

As a result of the increased attention that has been directed at technology transfer, the requirement for TTOs to have a clear structure and reporting relationships has come to the fore. Shattock (2001) argues that, because TTOs have become central to HEI success, and as a consequence of the extent of HEI-business collaboration, specialist offices have developed that need both co-ordination and leadership at pro-vice chancellor, vice-rector or vice president levels. Extensive research by Bercovitz *et al.* (2001) found that the structure and form taken by the TTO can impact the quality and number of technology transfer outcomes. The growth and importance in the importance of the activities of TTOs necessitates a prominent position in the HEI's structure and among the HEI's management team. This confers legitimacy on technology transfer activities and helps them to be viewed positively and accepted by academics. While there is no optimal structure that a TTO should take, a number of guidelines taken from best practice can facilitate the decision process (**Table 7.3**).

Executive Committees

The TTO should be an inherent part of a broader HEI structure that supports and fosters technology transfer. Relations with research offices are crucial, so that the TTO maintains an awareness of the research capabilities of the HEI. For larger agreements with particular IPR or financial significance, evidence suggests that TTOs may be supported in its decisions by an executive committee. This committee may be made up of senior members of staff and the top management of the HEI. The function of the committee is to provide direction on policy, act as a check on TTO activities, thus ensuring a balance of research exploitation and academic freedom, while at the same time buffering the TTO in terms of providing support for key decisions.

Indeed, in his review of best practice, Allan (2001) noted that large agreements may require regular meetings among senior management of the company and HEI, as well as regular exchange at the level of the TTO.

Table 7.3: Guidelines for Developing an Appropriate Organisational Structure for TTOs

♦ Ensure clear lines of reporting and responsibility.

♦ Ensure close interaction between commercialisation and research activity.

♦ Typically, the TTO director reports to a vice-president, vice-provost or vice-chancellor for research (Siegel *et al.*, 2003).

♦ Improve communication and teamwork among HEI personnel, and consider co-location of key offices.

♦ Co-ordinate the efforts of various offices to support researchers.

♦ A matrix-type structure tends to be most appropriate, because of its overall capacity to co-ordinate and incentivise the HEI-wide interface with business.

♦ The TTO must have a prominent position in the HEI structure and among top management.

♦ Executive Committee to oversee commercialisation activities.

♦ Development of specialists (by ILO activity or by research area).

♦ Although the functions are of research and technology transfer are related, the two offices have very different goals and should be administratively independent.

Source: Bercovitz et al. *(2001).*

Co-Location of Offices

To support the TTO further, some HEIs have encouraged team-work by co-locating related HEI offices. For example, about 10 years ago, Penn State decided to cluster the administrative activities that engaged industry into one facility. This has fostered co-operation, rather than competition, for establishing relationships with companies and sharing information. Some HEIs have gone so far as to incorporate their Office of Industry Research Relations and Office of Technology Transfer into a combined Office of Technology Transfer & Industry Research – for example, North Carolina State.

Degree of Autonomy

The degree of centralisation of the TTO, generally, will reflect both the culture prevailing at the institution and the size of the institution (Martin, 2000). A decentralised approach is natural in the case of a multi-campus HEI distributed over a number of locations – for example, University of Sao Paulo (Brazil) – in contrast to the centralised approach adopted by YISSUM at the Hebrew University of Jerusalem. Although YISSUM is a separate legal entity, it is subject to strict obligations as laid down by HEI policy.

Structure by Specialism

The *ad hoc* development of many TTOs has resulted in them handling a wide range of activities with limited resources. Offices generally cater for a number of different functions, ranging from promotion of the institution to external parties, research funding information, organisation of industrial placements and detection of academic entrepreneurship on campus. The trend in the US has been that TTOs have expanded from one-person operations to offices with several specialists working under a director. Each of these specialists tends to focus on one single aspect of technology transfer. This trend reflects how the role of the TTO has changed significantly over time.

Figure 7.3: TTO Structure: Specialist Activities

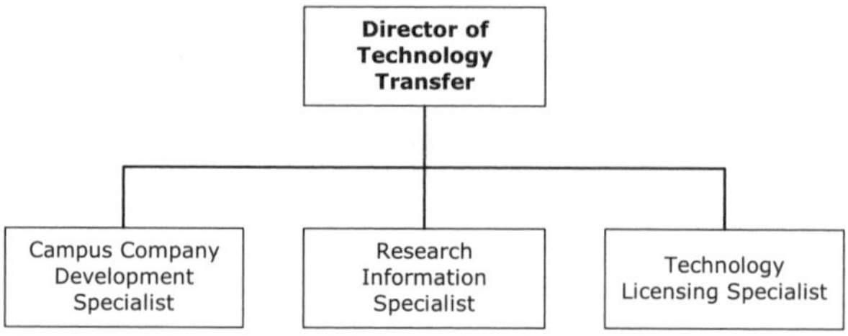

Another model evident from best practice institutions is that TTOs structure their activities in relation to the identified strategic research priorities or technology platforms of the institution (see **Key Issue 2**). This structure involves specialists in various research areas conducting the full range of technology transfer activities. This 'cradle to grave' approach has been termed 'case management'. The case-management style is commonly practised in many of the TLOs studied by Allan (2001) in his review of best practice (see **Figure 7.4**). In this system, one person is responsible for the actions required for a particular case, from disclosure through patenting, and sometimes beyond. This approach offers the advantage of centralising awareness and co-ordination of all major aspects related to a particular intellectual property. However, this management style requires the talents of very skilled individuals with experience in processes ranging from invention disclosure through to commercialisation. The breadth and depth of these processes usually leaves little time for proactively promoting the opportunity.

Figure 7.4: TTO Structure: The Case Management Approach

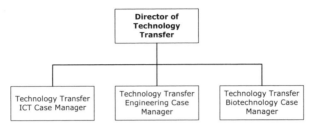

In general, both structures suggest a move from multiple-role 'one-stop entities' towards a more specialised focus, reflecting the technology transfer objectives of the HEI. These structures engender specialists and dedicated attention to specific aspects of the commercialisation process or research areas, so that hubs of expertise can be built up. The ultimate manifestation of such an approach would be a move towards the matrix structure of Duke University, as discussed in **Chapter 2** (see **Figure 2.4**).

In the context of such changes, the title of the TTO must reflect the role of the office and its positioning in the organisational structure. Many offices will have to adapt their structure to fulfil the pre-requisite for efficient and successful collaborations with industry. Graff *et al.* (2001) neatly depict stereotypes of the type of change that maybe required in **Figure 7.5**.

Figure 7.5: Structural Changes Required in TTOs

Structural Options to Overcome Resource Constraints

Best practice in the area of technology transfer usually assumes staffing and resources as a given. The reality for many smaller (especially non-US based) TTOs is that constraints will influence the structure adopted. As a result, the role of many TTOs has been generally re-active when it comes to research commercialisation. Evidence from best practice, however, shows that successful technology transfer function stems from a pro-active approach. This approach involves auditing and reviewing research programmes for projects with potential, protecting the IP generated from these projects and subsequently marketing them to industry. In a large number of cases, however, resource deficiencies, particularly legislative issues, constrain the ability of TTOs to operate to their full potential. As a result of such deficiencies, TTO involvement is often limited to serving as an information-point on a wide range of areas relating to external relations and assistance is curtailed to advice rather than expertise. Most TTOs are heavily constrained by budget and resource allocation and, therefore, should search for innovative mechanisms and creative solutions to overcome such deficiencies. Developing a meso-level structure may facilitate co-operation among TTOs, thereby reducing resource and expert deficiencies and leveraging economies of scale.

Key Success Factor 4:
Meso level structures (regional or national) may facilitate co-operation among TTOs, thereby reducing resource and expert deficiencies and leveraging economies of scale.

Regional Co-operation: Leveraging Resources & Expertise

In the context of deficiency in expertise, particularly in aspects of commercialisation and knowledge of the patent process, it may be beneficial to develop more co-operative regional or national mechanisms between TTOs. Such co-operative models may be a natural progression, given that research has suggested that there appears be a minimum efficient size for running commercialisation activities within HEIs. MIT's revenues are still only 3% of its research income, while a NHS report, for example, estimates that average revenues from technology transfer at leading US and UK HEIs are 2.5% of their research income (NHS, 1998). The same report then estimates that R&D expenditure of £20m per year is necessary for critical mass in technology transfer, that is, to cover the costs of a professional office. Lambert (2003) applied this data to the UK HEI sector, and noted less than 25% of HEIs would meet this threshold, even though 80% are now trying to run their own operations. This may explain the limited diffusion of best practice from TTOs in the US, as best practice writings have been monotonous in their prescription of large institution solutions and models.

Some have argued that partnership between TTOs may provide the essential means for undertaking work that is becoming evermore complex and interdisciplinary (Brint, 2005; Jankowski, 1999). The requirement for qualified specialist intermediaries is compounded by the introduction in 2004 of further EU regulation on technology transfer, adding further to the complex process of licensing and spin-outs (Forfás, 2004a: 32). Indeed, the EU puts special emphasis on collaborative arrangements between HEIs (Shattock, 2001).

One of the main barriers to commercialisation of HEI intellectual property cited in **Chapter 5** was the variable quality of TTOs. Clearly, for most TTOs, it is not feasible to staff their technology transfer functions with specialists who can work across a number of areas. While most HEIs run their own technology transfer operations, the reality is that only an elite few have sufficient levels of research activity to justify the investment required to build high-quality offices with substantial expertise. However, it may be feasible to have specialists in particular areas, if there was a system in place that allowed the technology transfer specialists to work across different HEIs. Variants of such co-operative approaches are not uncommon. In Denmark, Larson (2000) reports that, in the Copenhagen area, four HEIs have formed an amalgamation of biotech competencies in a type of joint property, also open to participation from private firms. On the same site, there is construction of a research park as an integrated part of the merger. The University of California model also highlights how this can be done. Such a system requires a high degree of co-ordination, with a senior manager taking overall responsibility for its operation.

Graff *et al.* (2001: 28) note 'it is reasonable to assume that the marketing of HEI research would benefit from economies of scale, meaning that, at least to an extent, the larger the represented research base, and thus the larger the portfolio of new technologies made available to license, the lower the marketing costs and the higher the probability of successfully selling licenses'. These conditions could be achieved if several HEIs used a single TTO to market their innovations, a move that would increase the variety and depth of technologies available and thus increase the attractiveness of the combined research base to potential business customers. The establishment of a middle-man organisation, therefore, can be a 'viable alternative'. Meyer (2003) notes that 'for campuses with a technology transfer office but with insufficient staff, it is possible to outsource some of the activities – licensing professionals assisting the HEIs with invention evaluation for IP and commercial potential, as well as marketing of inventions and license negotiation'. A small office may therefore want to share resources rather than hire second-tier talent (Lopatin, 2004). The Triangle Universities Licensing Consortium (TULCO) approach, in which one centralised office addresses technology transfer for several HEIs, is one such example. This was set up in Louisiana in the mid-1980s but dissolved a decade later when the universities established their own on-campus fully-self-sufficient technology transfer offices (Meyer, 2003).

A Co-operative Model for TTOs: Shared Services

The potential benefit of such a system of co-operation is two-fold. First, it provides an on-the-ground presence in each HEI for technology transfer issues. Stimulants for technology transfer identified in **Chapter 5** included good relationships with researchers and academics, coupled with knowledge of the research activities of an institution. Such knowledge is crucial in developing trust with academics and for promoting the merits of the commercialisation process. Second, particularly important in the context of the aforementioned expertise deficiencies, the model provides for specialist expertise in particular fields and industries to be shared. This is particularly salient, given that the lack of human resources with the right skill-mix is a major barrier to successful commercialisation (Rank, 1999). Such an approach is especially beneficial for less research-intensive campuses without an on-campus TTO, where a system-level person could be designated to serve several campuses. The model would also enable HEIs to share costs in terms of the legislative details of the patenting process. Indeed, if HEIs co-operate, they would have greater bargaining power for negotiating deals with the law firms. And, as well as providing economies of scale, co-operative mechanisms would provide scope for mutual learning and diffusion of best practice.

Table 7.4: A Co-operative Model of TTOs

A Co-operative Model of ILOs	
Shared Services	**HEI-based Services**
Licensing negotiation Market research for new technologies IP marketing IP management Spin-out creation Legal expertise	Raising awareness of IP issues Negotiating collaborative research contracts Reach-out to business Consultancy contracts Local contacts

Source: Adapted from Lambert (2003: 56).

A shared service model does not involve the closing-down of TTOs in any HEIs (technology transfer staff need close links with researchers to gain their support for commercialisation, to find out about new technologies in the research pipeline, and to locate potential industry customers in their specialist field) but rather it highlights the benefits of a *complementary* structure at a more macro level, drawing on the joint resources of a number of institutions. The Lambert Report (2003) provides some useful insights into the potential mechanisms and structure of such a co-operative body (see **Figure 7.5**).

Need for Government Support

In order for such a model to be a viable and sustainable option, government support is required to encourage the development of shared services in technology transfer on a regional basis. The expertise and legislative capabilities necessary to co-ordinate efforts may be readily available from government sources or funding agencies. Co-operative models suggest an enhanced role for the development agencies in facilitating business-HEI links.

The Belgian government, for example, has created a supportive context for technology transfer by sponsoring the establishment of a one-stop intellectual property centre/network, Belgium's Interuniversity Institute for Biotechnology (VIB), to serve the smaller public research organisations that lack the resource or critical mass to build their own fully-operational TTOs. Another example cited by the OECD (2003) comes from Denmark where, as part the implementation of new legislation, a grant of €8 million was set aside for the period 2000-03 to help HEIs protect and market their inventions. The grant helps HEIs to cover the external costs of patenting and marketing (up to €20,000 per invention), while also helping them to establish joint TTOs or networks along geographical/sectoral lines. **Table 7.6** provides some useful guidelines in relation to the creation of such a co-operative structure.[26]

The broadening of the range of issues confronted by TTOs clearly creates a challenge. More researchers becoming are aware of, and are motivated by, entrepreneurial incentives. Also issues such as company formation form a relatively new challenge for TTOs, which bring to bear a number of complex issues that may not have been anticipated (Allan, 2001). In such circumstances, TTO staff are stretched. These pressures, coupled with the difficulty of attracting and compensating employees with talents to master all such relevant areas, suggest that due consideration should be given to collaborative structures as suggested above. In an Irish context, ICT Ireland (2004a) proposed that

---

[26] Suggestions drawn from Lambert, 2003; Allan, 2001; Scott *et al.*, 2001, with additional insights from primary and secondary data sources.

an ICT intellectual property commercialisation centre be established with a €10 million annual patent fund for industry/HEI outputs. This would provide a 'one-stop-shop' for companies seeking to collaborate with third level research.

Table 7.5: Guidelines for Developing Co-operative Regional/National Structures

♦ The most research-intensive HEIs should be involved, where possible, to build on existing expertise.

♦ Initiatives should be supported financially by government agencies, which may also provide expertise to help manage and co-ordinate the structure.

♦ Non-prescriptive – HEIs in each region should agree themselves how to set up and shape the services, and the role that each institution should play.

♦ This voluntarism, however, will have to be accompanied by strict policies regarding the conduct of member institutions, particularly in terms the ownership of intellectual property and the confidentiality of information.

♦ Most knowledge transfer services should be kept in the HEI, including contract negotiation for consultancy and collaborative research and reach-out to business.

♦ It is crucial that some technology transfer staff should remain on-site at the institution to maintain and develop relationships with academics who wish to commercialise their research. Such first-hand contact is imperative in building trust and promoting commercialisation activities.

♦ Development agencies should support the HEIs in delivering the shared services.

♦ The shared service body should use its bargaining power to negotiate better access to legal and patent expertise.

KEY ISSUE 4: STAFFING, SKILLS & RESOURCES

A major challenge for TTOs is sourcing skilled personnel. Deals have become more complex and staff must master many competencies. Apart from solid knowledge in specific scientific fields, TTO staff must have legal and economic competence to judge whether inventions are patentable, marketing and business skills in order to find commercial partners and finally, negotiation and social skills to be able to finalise a good agreement (Karlsson, 2004). TTO activity, therefore, requires both technical and boundary spanning skills (see earlier). Overall, technology transfer staff need close links with researchers to gain their support for commercialisation, to find out about new technologies in the research pipeline, and to locate potential industry customers in their specialist field. Increased emphasis has been placed on marketing and negotiation skills as TTO staff can be successful in eliciting an invention disclosure and patenting it, but then suffer difficulties in licensing it to business. Barriers to effective technology transfer are often a function of the deficiency in skills and abilities of the TTO staff-member for co-ordinating activities and commercialising research (Graff *et al.*, 2001; Sanchez & Tejedor, 1995; Scott *et al.*, 2001). The varying demands placed on TTO staff highlights the requirement for strategic staffing, based on expertise and the need for continuous training and development (Lambert, 2003). **Table 7.7** provides useful guidelines to facilitate optimal recruitment of TTO staff.

Table 7.6: Guidelines to Facilitate Optimal Recruitment of TTO Staff

♦ Expect to pay top-rate professionals top-rate salaries.

♦ Do not try to grow someone into the technology transfer position. Use seasoned professionals only.

♦ Make sure the leader is both mentor and teacher.

♦ Hire those with an aptitude to understand science and the eagerness to put a business case around the science.

♦ Having staff with academic qualifications such as a PhD facilitates in gaining respect and managing relations with inventors.

♦ People working in the TTO must have the ability to make cold calls. The ability to get the deal done is a central requirement of the job.

♦ If possible, the Director of Technology Transfer should have solid experience in commercialisation and licensing, specifically in a HEI setting.

♦ Another preference is that the TTO should hire someone with experience of innovation in the region, and specifically with connections to venture capital and industry talent.

♦ Recruits should have the ability and people skills to foster a co-operative mind-set among researchers and commercial interests.

♦ Recruits should have the ability to work well with, and report to, other people within the HEI on progress and activities (for example, executive management committee, Offices of the Dean of Research, Vice President for External Affairs).

Source: Best practice and various TTO websites.

Strong Leadership

Clearly, having appropriately trained and skilled staff is an essential prerequisite to technology transfer initiatives. Best practice institutions go to great lengths and pay extremely good salaries to hire technology transfer talent. TTOs, therefore, require sufficient budget allocation to support these initiatives and to invest in commercialisation efforts.

The recruitment of appropriate and capable leadership is of crucial importance. For the University of Carolina, the leadership recruitment drive involved dedicating resources to hiring a leader with a technical background, experience in commercialising university technology, and the capability to foster a cooperative environment for innovation within the university.

Other leadership criteria drawn from best practice include:

♦ A strong record of leadership, management and administration of a sizeable unit, with demonstrated abilities to develop and articulate a vision for an organisation and to plan strategically.

♦ An in-depth knowledge of HEIs, the academic community, the investor community and industry, and of the corresponding cultures.

♦ In-depth knowledge and experience of intellectual property issues, the principles of licensing and spin-off company creation, as a means to carry out technology transfer objectives.

♦ Outstanding inter-personal skills and a proven collaborative management approach with the ability to consult widely, internally and externally, and to inspire confidence.

♦ A commitment to the broader goals of the HEI, and to building bridges with units of the HEI and between the HEI, industry and government.

♦ A strong record of mobilising resources and securing funds from a range of sources.

Research has shown that staffing within TTOs can help to explain why some HEIs are more proficient than others in managing intellectual property. According to Parker & Zilberman (1993), TTOs usually hire either a mix of scientists and lawyers or a mix of scientists and entrepreneurs/businessmen. In the former case, legal functions, such as the adjudication of disputes involving intellectual property rights and the negotiation of licensing agreements, are performed in-house. In the latter case, such functions are usually outsourced. Parker & Zilberman (1993) hypothesise that the entrepreneur/business model for TTOs may be more conducive to helping scientists form their own start-ups. It also seems reasonable to assume that TTOs staffed in this manner would be more effective in the marketing phase of HEI-industry technology transfer. Siegel *et al.* (2003) found that a substantial percentage of managers suggested that HEIs hire more licensing professionals with stronger marketing and business skills. Furthermore, in the entrepreneur/business model, legal expertise maybe shared among a number of TTOs. Siegel *et al.*'s (2003) research is again instructive, as their findings imply that spending more on external lawyers reduces the number of licensing agreements, but increases licensing revenue. The use of outside lawyers provides a signal that a HEI is aggressively protecting intellectual property.

Training & Professional Development

In order to keep TTO staff up-to-date with the patenting process, it is necessary to provide continuous training and professional development. In this respect, one of the important strengths of the US is the Association of University Technology Managers (AUTM), the national representative body for university technology managers. AUTM offers accredited courses that provide a range of training options for people involved in commercialisation. By supporting quality across the board, AUTM gives all HEIs the ability to raise their skills and experience. It also has strong industry involvement, with almost 50% of its membership coming from business. This helps improve HEIs' understanding of the needs of business and *vice versa*, through building networks for knowledge transfer.

In other cases, particularly in Scandinavia, governments tend to provide support to agencies to provide such professional development. In an Irish context, Enterprise Ireland has also initiated training activities, with the aim of increasing the levels of expertise and qualification for researchers and technology transfer personnel in the TTOs of universities and Institutes of Technology, including:

♦ The provision of expert mentoring to universities and Institutes of Technology through dedicated consultation with technology transfer staff from Columbia University, New York and Imperial College, London.

♦ An investigation of potential certificated IP and technology transfer education courses for research and TTO staff at universities and Institutes of Technology (Forfás, 2004a: 38).

EI has also funded a significant programme of R&D management training courses under its *Innovation Management* initiative, which has included modules on IP. While focused on industry participants, this initiative has been used in recent years by academic researchers and TT officers.

In terms of skills development and raising awareness, Singapore set up 'Technopreneurship 21', educating the domestic workforce through a series of school, HEI and workplace-based educational programmes. This has been matched by an active programme to draw entrepreneurs, managers and skilled workers from overseas (PACEC, 2003).

There are suggestions that skill deficiencies and cultural barriers to technology transfer maybe reduced through the creation of accepted standards for training and credentialing of HEI and industry technology transfer professionals. A focused effort on the part of the TTO to encourage such professional training and awarding of credentials may encourage business to interact with HEIs.

Development needs, therefore, must become central to the concerns of TTOs. Future skills analysis and development needs should be assessed and planned for.

Skilled & Flexible Personnel

Overall, technology transfer requires well-skilled and flexible personnel. Visible commitment from top management is essential in attracting the best staff to work in the TTO (Den Hertog *et al.*, 2003: 99). As technologies become more complex, and industry becomes more knowledge-driven, technology transfer specialists will become more important. In addition to qualifications in specific technological fields, greater weight must be given to the skills necessary to carry out the job. In the case of a technology transfer specialist, marketing, communications and negotiation skills are likely to be just as important as having an in-depth understanding of the science behind the innovation. Continuing involvement of the researcher may provide this scientific understanding. Most TTOs surveyed by the OECD stressed the importance of the TTO staff's informal relationships. This may suggest why, in many cases, it is recommended that staff hired for TTOs have industry experience and, hence, an understanding of industry but equally important a network of contacts and linkages (OECD, 2003).

Other mechanisms of staff development includes periodic staff meetings and formal and informal discussions of cases under management by TTO staff. At larger HEIs, such as Harvard and MIT, these war stories emphasise details that create an important internal knowledge-base. Sharing of this collective experience provides an array of how-to advice and extremely valuable insight, especially for junior members of the TTO teams.

Integration of the Strategy of the TTO & its HR Strategy

The human resource strategy of TTOs and the competencies they emphasise and nurture must be integrated and supportive of the objectives as laid out in the strategic plan. Stanford and the University of California, for example, focus strategically on the appropriate staffing and resource levels, and address policies devoted to recruiting and retaining staff, including their professional development. Consideration of, and investment in, the HR strategies of the TTO will serve as a further indicator of the commitment by the HEI and the TTO itself to technology transfer and commercialisation activities. Induction programmes in TTOs might involve placements in another TTO, particularly in the US, where incoming staff could get a feel for activities and see how fully developed offices operate.

Resources

Resources, particularly financial resources, to facilitate the process of technology transfer are essential for a more pro-active approach to technology transfer activities. In a large number of cases, however, TTO involvement is often limited to serving as an information-point on a wide range of areas relating to external relations and assistance is curtailed to advice and facilitation. Clear consideration of the skills required and matching of HR strategies with TTO objectives should aid in reducing such deficiencies.

Resource-sharing

In cases where such resource investments are not possible in-house, where they are constrained by the size or research base of the HEI, consideration should be given to structures that facilitate the sharing of expertise, especially expensive legal know-how. Marketing and business skills could also be obtained from interaction and involvement with business faculty members. Indeed, one of the most under-used resources, particularly in working with potential campus companies, is the facilities and expertise of the business or commerce faculty. Academics in commerce faculties have a potentially valuable contribution to make to spin-off company projects as part of the project work for suitable courses, as in the case of the MBA programme at the University of Oregon. The Chalmers School of Entrepreneurship programme, which allows post-graduate students to work on real-world cases in the form of spin-off enterprises, is also worthy of consideration. The benefits of such an approach have been highlighted previously.

KEY ISSUE 5: POLICY & PROCEDURES

In an Irish context, an ICT Ireland report (2004a) documented the need to develop the intellectual property management and commercialisation expertise and resources necessary to ensure effective and rapid exploitation of research generated in HEIs. The report noted that 'research institutions will need to ensure that the appropriate structures and resources for commercialisation activities are in place. Technology transfer offices will require resources to deliver the required services, and their personnel will need continuing professional development and training. Research funding bodies and others must provide appropriate support as the Irish commercialisation infrastructure develops' (2004: 29).

User-friendly Policy & Procedures

In order to encourage commercialisation activities, the TTO needs to develop concise, easy-to-understand transparent policies to ensure that procedures are maintained and followed. With respect to policies and practices, user-friendly means that they meet the norm of peer institutions and industry finds them attractive and wants to do business with the HEI. A key issue for many researchers, particularly young researchers, is a lack of understanding of intellectual property issues and channels for exploiting intellectual property. There is a need to develop user-friendly information packs to all researchers to help build their awareness. These could include literature, videos or CD-ROMs, or pocket-sized summaries of the commercialisation process and activities of the TTO.

A user-friendly approach could also involve:

♦ Providing adequate outreach/educational presentations and web material.

♦ Demonstrating the skill, ability and willingness to negotiate with all bodies.

♦ Being professional and factual in explaining why an invention will not be patented and/or marketed. At least 70% of the inventions received by the TTO will not be licensed. It is important to take the time to explain why the marketplace will not, or has not, decided to license an invention (Meyer, 2003: 11).

Other user-friendly approaches used in other institutions include the development of researcher kits for each type of agreement. These kits include: an explanation of when to use the agreement; a lay-language outline of the agreement terms; a research questionnaire to identify the relevant issues; and a step-by-step outline of the process. In order to ensure that policies are user-friendly the development process could involve a review or trial-run by a focus group of faculty researchers. Meyer (2003) notes that HEIs in the US have established advisory committees of HEI faculty and staff, peer institution technology transfer personnel, and 'friends of the HEI' from industry to ensure policies are user-friendly and meet the norm (Meyer, 2003). Policies should also be systematically reviewed to ensure that they are developed and/or updated at both the system and campus level, with each addition subsequently examined for user-friendly wording.

Disclosure Process

Evidence from best practice institutions shows that TTOs need to note and develop research with potential in a quick and efficient manner (Allan, 2001). Resources should be allocated so that appraisals are conducted quickly and there is a separation of 'highly promising' from 'promising' submissions. This helps establish legitimacy and the political value of the office.

Yale's Office of Co-operative Research is exemplary in terms of the disclosure procedure. The office operates a distinct strategy in order to capitalise on potential successes quickly. This involves:

♦ Establishing specific strategies to seek new inventions early.

♦ Examination of a large number of disclosure opportunities quickly.

♦ Reviewing and documenting probability of finding licensees through proven assessment techniques.

♦ Following up with a thorough review to select the strongest candidates for success.

♦ Ensuring that sufficient time is spent on the inventions deemed to have the most potential for success.

Knowledge of an efficient and fair system encourages researchers to disclose their inventions. The disclosure process in TTOs, therefore, should ensure a specified, rapid turn-around time for disclosure assessments, based on a clear understanding of the proposed technologies. At the University of British Columbia (UBC), approaches to this include the following:

♦ A rapid pre-disclosure process involving a meeting between a researcher and a UBC TTO member of staff, and/or a brief summary statement may be introduced. In many cases, this may yield immediate clarification regarding what steps, if any, are appropriate.

♦ An advisory committee or peer review process provides more informed input into disclosure assessments. One manifestation of this review may be a section of the disclosure whereby the researcher recommends the specific peer review process she/he considers appropriate.

♦ A short, specified response time, for example 30 days, may be established, after which the UBC TTO should either commit to pursuing a technology or return it to the researcher.

Essentially, evidence from best practice indicates the importance of timely responses to invention disclosures, communicating clearly with researchers and leveraging expertise to ensure the disclosures chosen are the ones with the most potential opportunity.

Negotiating Agreements

In addition to HEI licensing policies, premature definition and valuation of intellectual property can become an obstacle at the initiation stage of a collaborative project. Granting the company the right of first refusal to negotiate an exclusive license is one commonly-used practice to delay concrete negotiations until the commercial value of an invention is easier to assess. When the extent of commercialisation activities are of a sufficient magnitude, or in the case of larger HEI-industry research relationships, the use of master contracts or templates should be considered. Major HEIs in the US use such templates and make them freely available by posting them on websites. These templates provide useful parameters for negotiation. Furthermore, standardisation reduces the risk of error or litigation as a result of certain clauses. Such contracts should be continuously reviewed to ensure that they keep up with most recent developments. HEIs also should consider developing model agreements for single research projects and ensure that the terms do not unduly disadvantage small and medium-sized companies.

Confidentiality agreements, when necessary, should be signed by the company, the HEI, and the researchers involved. The company and the HEI must take responsibility for safeguarding confidential information. Publication delays to protect intellectual property rights should generally be no longer than 60 to 90 days. Any publication delays should be carefully monitored, both to preserve academic freedom and to protect against any early disclosure that might invalidate patent claims. Clear procedures with regard to this will help in easing discomforts that some academics naturally will have in terms of the commercialisation of their research results.

Conflicts of Interest

Conflict of interests as defined by the OECD (2002) refers to a situation where 'an official's private interests – not necessarily limited to his/her financial interests – and the official duties in his/her public function are in conflict (actual conflict of interest) could come into conflict or could reasonably appear to be in conflict (potential conflict of interest)'. Conflicts of interest can exist among faculty, between the TTO and researchers and also among non-faculty such as graduate students.

Even when expectations are well-managed, some form of conflict can become apparent. Financial conflicts of interest arise when scientists' private financial interests and their research converge in a way that might call into question their ability to make unbiased decisions relating to their work. Even the perception of possible conflicts of interest could prove to be extremely damaging to the reputation of the HEI and company concerned. Perceptions of a conflict of interest can weaken public trust – a particular concern for research HEIs, which heavily depend on government research funding.

Conflicts of commitment are generally defined as anything that might interfere with a faculty member's full-time duties. Many HEIs have formal policies limiting the amount of time that a faculty member can spend engaged in outside activities (for example, MIT).

A natural outgrowth of an increased focus on HEI-business interaction is an increased likelihood of conflicts of interests arising. The real issue then becomes what methods should be put in place to resolve conflicts once they come about, since their emergence is a predictable part of the process of change (Etzkowitz & Leydesdorff, 1999: 114).

TTOs must develop clear guidelines to deal with conflicts of interests as they arise. Guidelines provide templates on how to avoid or manage potential conflict of interests between the researcher's obligation and the more entrepreneurial activities such as patenting inventions, carrying out contract research or working in a start-up company or spin-off. Guidelines for dealing with individual conflicts of interest vary but there are common elements, such as disclosure of all relevant financial interests and activities outside the framework of employment or disclosure of financial interests to non-government sponsored research. Among OECD countries, only Denmark and Germany have developed national guidelines concerning conflicts of interests involving research staff and intellectual property activities. In other countries, such as Canada, France, Ireland, Holland, UK, and the US, guidelines are developed at the level of the institution or funding agency.

As HEI officials, researchers, and the companies with which they collaborate study these conflict-of-interest issues, there are a number of basic principles that serve as parameters to the creation of guidelines, including:

♦ The core values of academic freedom must be maintained.

♦ Industry funding cannot, and should not, be viewed as a substitute for adequate, long-term public financing of basic scientific research.

♦ HEIs and companies should seek transparency, clarity, and consistency.

The precise mechanisms for managing a conflict usually depend on the details of each case. Options can include divesting troublesome assets, ending consulting arrangements, withdrawing the researcher from the project, independent review, and disclosing significant financial assets in any published report on the research. A useful strategy for preventing potential conflicts involves ongoing education, aimed both at faculty members and at graduate students who hope to become practicing scientists (BHEF, 2001). Research HEIs tend to institute rather strict policies relating to the conflict of interests, in particular with regard to spin-offs (Hernes & Martin, 2000: 35).

Sponsored Research

Often conflicts of interests arise as companies request background rights to sponsored research projects. Companies have legitimate reasons for requesting background rights to sponsored projects and, as part of their due diligence, should assist HEIs in locating potential conflicts. HEIs have legitimate reasons for not providing background rights, but they should make a strong effort to do so when appropriate and feasible. HEIs should closely consult with faculty and confirm that all contractual obligations can be met before signing binding agreements (BHEF, 2001). Detailed consideration and consultation can reduce the likelihood of conflict and facilitate in resolving it, should it arise.

Other conflicts of interest may arise with regard to the misuse of student time. For example, when a faculty member holds an equity stake in a company that sponsors HEI research and has graduate students working on that research, tensions and suspicions can arise. In order to address this issue, experienced HEIs have developed policies to deal realistically with these issues. Although some faculty members may wish to minimise involvement in collaborations by the HEI's research administration and departments, others assert that open communication with these bodies can prevent abuses, as well as suspicion and misunderstanding about such arrangements. Some HEIs and departments simply do not allow students to become regular or part-time employees of research sponsors.

Intellectual Property Rights

The activities of TTOs are greatly facilitated by active protection and management of intellectual property. Due consideration to IP-related issues can simplify negotiations. Policies that help with this include those shown in **Table 7.7**.

Clear Codified Intellectual Property Rights

One of the barriers to commercialisation cited in **Chapter 5** was a lack of clarification over intellectual property rights. Such confusion can result from a lack of transparency over procedures and regulations. Clearly, when establishing collaborative research partnerships, it is important to determine at the outset the ownership and exploitation rights for any IP that may be generated. Business and HEIs both report that negotiations on the terms and conditions of IP ownership and exploitation can be extremely lengthy and costly. Research continuously refers to the importance of clear codified guidelines, particularly in the areas of ownership of intellectual property, equity shares and distribution of royalties.

Table 7.7: Intellectual Property Protocol: Main Features

- The common starting point for negotiations on research collaboration terms should be that HEIs own any resulting IP, with industry free to negotiate licence terms to exploit it.
- But, if industry makes a significant contribution, it could own the IP.
- Whoever owns the IP, the following conditions need to be met:
  - ◊ The HEI is not restricted in its future research capability.
  - ◊ All applications of the IP are developed by the company in a timely manner.
  - ◊ The substantive results of the research are published within an agreed period.
- On all other terms, the protocol should recommend flexibility where possible to help ensure that the deal is completed.
- The Funding Councils and Research Councils should require HEIs to apply the protocol in research collaborations involving funding from any of the Councils.

Harmonisation of Intellectual Property Rights

TTOs should be involved in efforts to encourage further harmonisation – or at least compatibility – of national rules regarding IPR, which may also in turn facilitate collaborative research by reducing transaction costs. In many countries, government agencies have been heavily involved in promoting such harmonisation and diffusing knowledge and expertise in respect of IPR, in the form of training programmes and workshops dealing with IP.

In an Irish context, Cronin (2004) has argued that a key factor in facilitating Ireland's future competitiveness will be the efforts at establishing and diffusion of a code of practice for the management of intellectual property in 100% publicly-funded research. Policy safeguards, such as those specified by funding bodies and agencies, can help balance the research and commercial goals of institutions. Individual TTO should also set its own judicious guidelines. The work by the Irish Council for Science, Technology & Innovation (ICSTI) in developing a *National Code of Practice for the Management & Commercialisation of Intellectual Property from Publicly-funded Research* is an important first step ensuring that explicit regard is given to exploitation of research results in the policies of research institutions. The implementation of the procedures outlined in the Code should be a priority for all public research organisations (ICT Ireland, 2004).

Distribution of Intellectual Property Benefits

Guidelines should clearly outline the reward system of shared benefits. Ideally, rewards should be weighted positively towards the inventor, to act as an incentive for research commercialisation. This is particularly the case for HEIs trying to develop and attract attention and interest towards commercialisation activities. Elements of such policies should include the researcher benefiting financially at an early stage. Northwestern University's policy allows for 20% of the net income to be transferred to a research account of the inventor's choice. This should prove an incentive for departments to encourage technology transfer *via* IPR. Clearly, it is critical to have distribution policies in place prior to beginning the commercialisation process and before the research concept becomes valuable, at which stage reaching an agreement can become a very complicated and legal business. The most successful policies are presented in a user-friendly format and described at orientation programmes. In addition, IPR policies should reflect the fact that innovations are often developed as a result of a team effort and, thus, benefits should be shared between members of the team.

Perceived Publishing Constraints

Companies should have secure rights to the IP they want to commercialise, but it is also important that any deal on IP should not unreasonably constrain the HEI from publishing the results in a timely fashion, from doing further research in the same area, or from developing other applications of the same IP in different fields of use. It follows from these points that there should be as much flexibility as possible in the distribution of IP rights between HEIs and business.

Flexibility in Terms of Various Intellectual Property Policies

It has been noted that the maximum creative use of IP allows the full economic potential of research collaboration to be unlocked (Sanchez & Tejedor, 1995). If business negotiates full ownership of IP with strong restrictions on HEI use, this may reduce the total economic impact of the IP in the future. Therefore, it is important that there is flexibility in the distribution of IP rights between HEIs and businesses. Increased flexibility in the use of IP prevents it from being locked up in a way that limits its exploitation across as wide a range of areas as possible. Research on best practice suggests the best way to meet these objectives is to introduce an IP protocol (Allan, 2001; Lambert, 2003) (see **Table 7.7**).

Joint Research Contracts & IP

It is much more difficult to agree the ownership of IP in research projects that have been funded by both HEIs and industry. Most business funding for HEI research is in this form. IP ownership is often strongly contested in these research collaborations, as the sponsors have different interests in the rights to exploit and use the IP. The rationale guiding HEI demands for the IP is that ownership is required to ensure that their future research is not held back. Industry often argues that it needs ownership to protect the investment that will be required to develop the IP into a commercial product. Although ownership and control of intellectual property resulting from a collaboration must be decided by the collaboration partners, it usually will be appropriate for the HEI to retain ownership. Both parties should remain flexible during negotiations, and the key measure should be whether the corporate partner has the ability to commercialise the fruits of the research to the benefit of the public. HEIs should update their copyright policies to allow industry sponsors to be granted licensing terms on a basis similar to that provided with patents. Occasionally, the resolution mechanism is that IP ownership should be proportionate: the party that makes the largest contribution (intellectual as well as financial) should have first rights on the IP.

Property negotiations, however, are also frequently hampered by HEIs taking a restrictive approach to licensing and placing too high a value on their intellectual property contributions. Research by Siegel *et al.* (2003) and Scott *et al.* (2001) found that some HEI boards of trustees may see technology transfer activities more as a revenue source than as a component of the HEI's public responsibility to assist in commercialising research results. This attitude can raise barriers to negotiations that actually reduce revenue in the long term. Given that only a small percentage of HEI-generated inventions produce significant revenue, some participants likened the strong emphasis on protecting proprietary rights of some HEIs to 'buying lottery tickets'. HEIs, therefore, would benefit from being less aggressive in exercising their IP rights (Siegel *et al.*, 2003). More commonly, problems arise with projects with less well-defined outcomes than clinical trials. Several participants noted that HEIs should avoid overselling in terms of potential accomplishments and timelines.

Clearly, the role of intellectual property in the innovation process varies by field. In some cases, HEIs do not seek patents on their inventions unless an industry licensee has been identified. Allan (2001) notes that this approach is much more likely to facilitate commercialisation rather than a blanket policy of not patenting inventions outside the life sciences, which is evident at some HEIs. **Table 7.8** offers principles that serve as useful guidelines for TTOs in managing IP.[27]

---

[27]   Adapted from ICT Ireland (2004a); ESG (2004).

Table 7.8: Intellectual Property Rights Guidelines

♦ The cost of use of rights should be very industry-friendly to encourage take-up.

♦ HEIs should have a mandate to ensure take-up of their IP, and not a narrow mandate to make money. A HEI's inability to achieve this take-up should reflect badly on it; and should be looked into on an annual basis by an authority that can take action to address this problem.

♦ Great care should be taken to ensure the right balance between selling IP too cheaply and not selling it at all because of overpricing. Over-aggressive exercising of IP rights is a common problem among HEIs (Siegel *et al.*, 2003).

♦ An institution should not aim to build up a large amount of IP that it tries to license. While seemingly attractive, few companies are making money on selling IP (in contrast to selling products based on them). World-wide, few applied research institutions or HEIs make any substantial money from IP, and most make none at all.

♦ The monetary value of a HEI should be measured on how it improves industry, and provides highly experienced researchers, and not on the money it makes from selling IP (see evaluation below).

♦ Nevertheless, incentives are important at all stages from research to successful productisation, and institutions should have a financial incentive when selling IP.

♦ The IP rights should depend to a small degree on whether the research is collaborative or contract:

◊ **Collaborative:** The research results should be openly published, with widely-held rights to use. Tightly restricting the IP rights to within a HEI will lead to two very negative results: rare, if any, exploitation; lack of industrial involvement.

◊ **Contract:** HEIs should never compete with consultancy companies by carrying out work that is owned by the funding company. Results should be openly published, perhaps with early access for the funding company.

♦ IP rights need to be very clear, so that no time is wasted in negotiation.

Keeping Costs in Check

It is crucial for TTOs to keep budgets under control and have strict financial assessments of their activities. This is particularly the case in terms of legal fees, the cost of commercialisation and the time lag of any recoveries as a result of successful commercialisation activities. To facilitate with this process, some HEIs – for example, John Hopkins – have adopted a software database (*FileMaker Pro*) to track all commercialisation activity for a technology. This includes information pertaining to ownership, assessment, marketing, licensing, royalty income, and legal fees. Legal fees are checked regularly and are contested if the TTO finds them unreasonable.

Rebilling

Another method of managing the legal fees is a practice called 're-billing'. Licensees are billed for all legal work associated with the technology licensed, including patent execution costs. While not all licensees pay 100% of the costs, willingness to do so is a factor in license negotiations. Other methods of recording financial activity include documenting the level of funding and other support provided to the TTO and comparing it to peer HEIs. The approximate number of patents applied for and granted in a year, as well as the total number of patents granted still under management, is important in estimating the TTO's potential costs and revenues.

Indirect Costs

In terms of sponsored research, it has been noted that industry may seek to stretch resources by not paying indirect costs and faculty may pressure the HEI to agree. Given the substantial indirect cost rates of research at HEIs and elsewhere, it is understandable that some companies are reluctant to pay, or wish that all the funding be used only for research. If indirect costs are waived on industry research, however, they must be made up somewhere else (tuition or other research grants). Indirect costs are a legitimate expense of performing HEI research. In most cases, companies should expect to pay at least the negotiated facilities and administrative charge for the research they sponsor in HEIs. As part of cost management procedures, all HEIs should issue a statement on indirect cost practices in HEI-industry research.

Several HEIs have introduced creative incentives to encourage industry to pay indirect costs. For example, some HEIs have traded current overhead recovery for a greater share of downstream royalty income or for equity. Care must be taken, however, because indirect costs are current and downstream income is uncertain, and any trade-off must not short-change another part of the HEI. Another example is the State of California's Micro-Electronics Innovation & Computer Research Opportunities programme, launched in 1981. For approved projects, California puts up one-third of the funding, the company puts up one-third, and the University of California campus involved waives overhead on the industry and state funding, essentially providing another third. HEI participants reported that this approach has been very successful in encouraging a broader range of companies to support research.

Key Issue 6: Mechanisms for Technology Transfer

It is evident from the proceeding discussion that, in the context of their strategic plans, TTOs should pursue a focused and planned approach to commercialisation and technology transfer. Attempts should be made to leverage research and resources from previously identified technology platforms. Research in this area stresses that TTOs and their activities are distinct products of their associated institutions history and development path. Each TTO, therefore, must choose a mechanism for technology transfer that suits their specific needs and circumstances.

Key Success Factor 5:
There is no one best mechanism for technology transfer.
Each TTO must choose mechanisms for technology
transfer specific to its individual needs and circumstances

Spin-offs & Start-ups

Spin-offs and start-ups are seen to be a direct manifestation of technology transfer and, therefore, prove popular technology transfer mechanisms for HEIs initiating or developing mechanisms for technology transfer.

Elements of good practice for developing spin-offs include:

♦ Provide clear and transparent guidelines to all HEI staff wishing to set up a spin-off company, so that the prospective entrepreneurs have a solid framework before they make a judgement about whether to proceed with setting up the company and how this should be done.

♦ Pro-actively provide information, contacts and support on how to establish a company.

♦ If the HEI owns the IP that protects the innovation or technology which forms the basis for the company, jointly discuss and decide at the outset with the academic inventors or prospective entrepreneurs whether setting up a company is the best option.

♦ Allow prospective entrepreneurs to remain in part-time research or teaching with the HEI, if they so desire; there should be clear demarcation of responsibilities and practices between the two jobs.

♦ Provide advice to prospective entrepreneurs on what is required to operate and run the company effectively and what additional expertise may be needed to be recruited to fill the gaps.

♦ Help develop business plans to define clear commercial goals for the new company, as well as addressing market research and sales strategies.

As documented in **Chapter 4**, it is important to have appropriate support mechanisms in terms of allowing academic staff leave to work on such ventures. Starting a new firm does not necessarily imply that they leave their academic position and take up a consulting position or a board position in the new firm. In order to operate company formation activities successfully, HEIs must have the appropriate level of management skill and expertise, sufficient resources as well as clear contracts documenting responsibility.

Some HEIs have diverted attention away from company formation activities. MIT's Technology Licensing Office licenses inventions as non-exclusive or exclusive licences to industry and local venture capital firms rather than getting involved in the complexities of spin-out formation. The TLO provides a shop window for industry to view its intellectual property and agrees as many licence deals as possible.

The Risks of Spin-outs

A start-up is only one of the exploitation mechanisms at a TTO's disposal for obtaining maximum benefits from its research results. The Lambert Report (2003) noted that high spin-out rates come at a cost to licensing. While spin-out activities of TTOs have led to some successful companies being created, there are signs that the pendulum has swung too far and that too many spin-outs are now being created, some of low quality.

TTOs, therefore, need to assess carefully the likelihood of a spin-off being successful and assess the resources and time that can be reasonably invested in encouraging such activities. While an attractive method of technology transfer, often resource constraints and a lack of expertise makes these activities unsustainable. This is best captured by the OECD (2003: 99) findings: 'in most cases, for commercial or legal reasons, the Industrial Liaison Office will have to consider more traditional technology transfer routes, like co-operative research or licensing, to moderate the potential risk.'

Incentive Structures for Spin-outs

In recent years, a number of HEIs have sought to implement innovative incentive structures by starting or expanding programmes aimed at facilitating the launch and growth of start-up companies based on HEI research. Such structures include HEI-managed incubator facilities, provision of seed funding in return for equity, and focused efforts to attract management talent and catalyse company formation.

A HEI moving in the opposite direction is the University of Arizona, which settled a patent suit brought by a start-up several years ago, and now licenses technology to established companies only.[28]

In respect of incubator facilities, it has been noted that TTOs need:

♦ To increase familiarity with business incubation issues and include them in the scope of policies and operations.

♦ Study regional success stories such as the Research Triangle Park and others.

♦ Solicit observations from other HEI-affiliated operations.

In the US, TTOs that use incubator facilities are often members of associations where they can leverage knowledge and communicate ideas. Such associations include the National Business Incubator Association and Association of University-related Research Parks. While, in some cases, the TTO may not manage the incubator unit or science park, there is usually some well-developed reporting structure to facilitate collaboration and communication between the incubator/science park and the TTO. In the UK, it is estimated there are now over 60 science parks, with an estimated 1,700 firms employing some 40,000 people (PACEC, 2003).

Training/Seminars for Prospective Academic Entrepreneurs

Where it has been decided that company formation should be a central objective of the TTO, TTOs, in conjunction with their respective business schools as well as development agencies in the region, should run training courses and information seminars in enterprise development for researchers who have decided to undertake commercialisation through a spin-off enterprise. There is a need to increase awareness of what is required for this process to provide researchers with the full information they need to make the right decisions.

Licensing

Licensing has become one of the most popular avenues for research commercialisation activities. Licensing is less resource-intensive than spinning out new companies – both in terms of people and funding – and has a higher probability of getting technology to market. It is often the quickest and most successful way of transferring IP to industry, and has the advantage of using existing business expertise rather than

---

[28] Blumenstyk (1995).

building this from scratch. It must be noted however that patents and other forms of IP should be taken out selectively, as not all results need to be appropriated or should remain in the public domain.

In order to successfully develop licensing capabilities, TTOs will need to have greater capacity and resources for managing IP. Critical in this respect is government support and funding to increase the availability of proof-of-concept funding. Proof-of-concept funding is used to establish whether a new technology is commercially viable or not. It is the first stage in transferring IP to the market. Proof-of-concept funding is needed for both licensing and spinning-out. The level of investment is normally up to €70,000 per invention. TTOs, therefore, should actively promote the importance of proof-of-concept funding and work with local development agencies and other TTOs to ensure adequate resources are allocated to these activities.

The advantage of using licensing strategies is that licensing can be used to maintain access to IP so that it is not lost, which is not the case, for example, should a spin-off company fail. Furthermore, licensing activities build on existing business expertise rather than attempting to build these up from scratch.

Portfolio Approach: Pursuing Many Deals

One approach used by US institutions has been termed 'portfolio management', whereby TTOs pursue a number of commercialisation channels to minimise the risk of being exposed by any particular one. Statistics from the Association of University Technology Managers (2002) established a positive correlation between royalty income and the number of active licenses at an academic institution. Accordingly, many institutions adopt a portfolio approach in order to pursue many opportunities simultaneously, aiming for success collectively. Linked with such an approach is the prioritisation of daily activities to make progress on signing deals. Such prioritisation should be made in the context of the TTOs' overall strategic objectives. Given that one in nine deals pays out royalties, the success of the TTO will depend heavily on the number of deals made. A portfolio approach serves to spread the risk of commercialisation activities. Based on priorities, as laid out in the strategic plan, and based on each HEI's specific contingencies, a number of different channels will be pursued to facilitate in getting the deal done. Such an approach is often more realistic in the case of resource constrained institutions.

Networking

The importance of networking in evaluating disclosures for potential, and in facilitating with commercialisation activities, cannot be underestimated. The staff at the TTO of Georgetown University frequently liaise with other institutions and bodies in assessing the commercial applicability of various technologies. Often, this networking takes place on a more formal basis at a regional level with existing regional enterprise development agencies. Examples of HEIs networking with existing regional enterprise development agencies were highlighted in Belgium (University of Louvain-La-Neuve) and in Finland (Technical University of Helsinki). A similar initiative is underway between Trinity College, Dublin and the Dublin Business Innovation Centre. Such co-operation with agencies in the region to access specialist advice and other types of support should be encouraged and HEIs should develop networks for this activity.

Other Mechanisms for Technology Transfer

Other mechanisms for technology transfer include sponsored or collaborative research, which is most favourable when the researcher continues research in the laboratory that is relevant to commercial endeavour (Jensen & Thursby, 2001). Collaborative research hedges the downside risk of lost academic opportunity, as the researchers' effort allocations are less constrained in such a grant agreement as opposed to if they were directly involved in the company as consultants, board members or founders.

KEY ISSUE 7: EVALUATING TECHNOLOGY TRANSFER

A crucial part of developing a more strategic approach to TTO activities involves the introduction of an evaluation mechanism to ensure that the office is conducting and prioritising activities as set out in its strategic objectives. As noted in the previous chapter, a critical, but often neglected, part of commercialisation and technology transfer procedures is their evaluation and measurement. Taking stock of organisational practices in HEI management will be useful in many regards. Given the embryonic nature of the TTOs, generally there is a need to document practices being conducted. Administrators often express a strong interest in benchmarking their IP management practices relative to peer institutions. Documentation of activities can also be disseminated to various stakeholders. In a HEI context, this will serve to legitimise and make transparent the role of the TTO as part of the general HEI structure.

Holistic Approach

We have emphasised that there is no one model for a TTO. Learning can be facilitated by diffusion of best practice and benchmarking on a number of quantifiable outputs, while also trying to continuously improve process activities. **Chapter 5** documented the quantitative variables used by the AUTM and various other studies to measure the output of TTOs. A more valid approach will focus holistically on efficiency and effectiveness and, therefore, capture processes and activities and the broader contribution to civil society by improving the quality of life and the effectiveness of services. For example, as an indicator, the revenues received from contract research could be complemented by the average size and length of the contracts to provide a sense of the depth of the research assignments. This should be the objective of any evaluation mechanism introduced to monitor the activities of TTOs. As Molas-Gollart *et al.* (2002: *iv*) comment: 'an approach that focuses purely on HEI commercial activities is likely to miss large and important parts of the picture'.

The introduction of measurement systems generally helps to encourage behaviour by defining activities that previously went unnoticed or were unusual. Often, actors will use the measurement system to guide their behaviour (Molas-Gollart *et al.*, 2002: 58). In this respect, measurement can form a perfect lever for the promotion, awareness and implementation of strategic objectives (Simons, 1994). **Chapter 5** documented different methods of looking at TTO activities, capturing not only output and outcome but equally important process-based evaluations.

Guidelines in relation to the development and use of indicators include:

- ♦ **Acknowledging 'variety of excellence':** Indicators need to be sensitive to disciplinary effects and avoid biases that may reward disciplines that exhibit the most visible direct link to commercialisation activities or successes.

- ♦ **Commercialisation indicators are not enough:** Indicators of HEI commercialisation are not a sufficient guide of TTO activity. Commercial activities are heavily concentrated in particular disciplines and the returns to commercial activities are highly skewed. On their own, commercialisation indicators are a poor reflection of the overall economic and social benefits of the HEI sector.

- ♦ **Use a variety of indicators:** There are no magic bullets in indicators of commercialisation activities. A variety of indicators need to be collected.

- ♦ **Some data is better than no data at all:** Resource and time constraints on TTOs often render detailed collection of data on indicators a futile mission. Following a rationale akin to Molas-Gollart *et al.* (2002), it must be appreciated that even small efforts in this area can be extremely beneficial.

Table 7.9 outlines some potential indicators that may be included as part of a measurement model of TTO activities.

When considering the potential use of new indicators, it is important to assess the effort necessary to collect, analyse, and update the data. It is common to underestimate the work that is needed to collect comprehensive data necessary to carry out proper evaluations and impact assessments. Labour-intensive techniques can be applied in one-off studies, but cannot be used as the basis of continued comprehensive studies over time. The use of indicators that are labour-intensive to gather, generates not only a problem of costs, but increases the administrative overhead on academic staff and can lead to increased centralisation of HEI research management. Other issues in relation to evaluation mechanisms are discussed below.

Continuous Review of Best Practice

It is important that research institutions, and TTOs in particular, continuously seek to operate best practice methods for increasing research commercialisation activity. This includes closely monitoring practices in other countries and should include overseas study trips specifically to review practices relating to technology transfer and research commercialisation. These reviews and fact-finding missions could be carried out on a collaborative basis. Furthermore, the horizontal collaboration between HEIs or meso-level structures referred to earlier should facilitate in communicating and diffusing best practice across a number of institutions

Table 7.9: Potential Indicators to Measure ILO Activities[29]

Indicator	Data collection instrument	Collection costs
General		
Technology commercialisation	TTOs may gather data at some HEIs. At other HEIs, such information may be held by central administration or at the departmental level.	Moderate
No. of patent applications		
No. of patents awarded		
No. licenses granted (including option agreements)		
Royalty income (including option fees)		
Median value of royalties (including option fees)		
Entrepreneurial activities		
No. of spin-offs created in the last 5 years	Turnover/profits from spin-offs and commercial TTOs may gather data at some HEIs. Elsewhere, information may be held by central administration or by departments.	Moderate
No of current employees in spin-offs created in the last 5 years		
Turnover/profits from spin-offs and commercial arms		
Development funds and loan facilities provided by HEIs to support start-ups		
Commercialisation and use of HEI facilities		
Income derived from leasing/letting/hiring of S&T HEI facilities (laboratories and testing facilities)	Data to be gathered from contracts, authorisation forms, or booking procedures. Some of the information will be available	Medium

29 Largely drawn from Molas-Gollart et al. (2002), with suggestions by Siegel et al. (2003), Graff et al. (2001) and the AUTM Survey 2002.

Indicator	Data collection instrument	Collection costs
Total no. of days spent by external (non-academic) visitors using laboratories and testing facilities	centrally and some at departmental level. Data collection process may be labour-intensive.	
Income derived from leasing/letting/hiring of cultural and HEI leisure facilities (theatres, conference rooms, sport centres, …)		
Total no. of events run and organised by the HEI for public benefit		
Income derived from leasing/letting/hiring of office and library space to industry and social groups		
Advisory work		
No of invitations to speak at non-academic conferences	Information could be collected as part of the annual appraisal process or through a survey.	Medium
No. of invitations to attend meetings of advisory committee of non-academic organisations		
Contract research with non-academic clients		
Value of contract research carried out by the HEI	Data to be gathered from information available in contracts and held centrally. Collection maybe labour-intensive.	Medium
No. of contract research deals (excluding follow-on deals) signed by the HEI with non-academic organisations		
Non-academic collaboration in academic research		
No. of refereed publications authored with non-academics	Information could be collected as part of the annual appraisal process or through a	Moderate

Indicator	Data collection instrument	Collection costs
No. of non-academic organisations collaborating in research projects	survey. Data to be gathered from information available in contracts and held centrally. Collection maybe labour-intensive.	
Value of contributions (in cash and kind) provided by non-academic collaborators		Medium
Flow of academic staff, scientists and technicians		
No. of faculty members taking on a temporary position in non-academic organisations	Data to be gathered from personnel records and CVs held centrally at HEI level. Collection process maybe labour intensive.	
No. of employees from non-academic organizations taking temporary teaching and/or research positions in HEIs		
Student placements	Data to be collected from central student records.	Medium
Learning activities		
Income received from non-credit-bearing teaching and associated activities (courses, collaborative learning) undertaken	Data may only be available at the departmental level, and could be gathered by setting up suitable mechanisms.	Medium
No. of different institutions that have attended or have taught in non-credit-bearing teaching and associated activities		
Social Networking	Information could be collected as part of the annual appraisal process or through a survey.	High

Annual Review/ Report

It is evident from the research conducted for this report that most best practice institutions produce annual reports of their activities. Annual reports or public documents noting evaluations may also enhance the reputation and credibility of the TTO among business. Such reports document the extent and level of technology transfer and licensing activities, as well as recent policy development and changes. These reports are made available on the TTO or HEI website and serve as a useful tool to communicate the role of the TTO to internal and external stakeholders, as well as promoting its professionalism. Reports also document success stories and provide details of the processes of commercialisation used in each of these cases. These accounts can be useful in altering mind-sets and attitudes to commercialisation.

For smaller institutions, an annual report may be too ambitious an undertaking; instead, the TTO should attempt to review systematically its operations and its effectiveness in carrying out its obligations to all stakeholders every two or three years. Publication of the methodology and results of this review, taking account of the stated objectives of the TTO, its performance against key performance indicators, evaluations of management and activities, should ensure transparency and communicate the activities of the TTO to all stakeholders.

There is obviously a necessity and benefit for TTOs to document and report the activities it is conducting and the outcomes of such activity. Increasingly, funding bodies are putting pressure on TTOs to develop such evidence. Commercialisation plans and abilities are often part of funding criteria and so it is beneficial to the HEI that the TTO can demonstrate professional and pro-activeness in this respect. Crucially, users should be involved in the design of evaluation measures, as this creates buy-in and awareness.

Another aspect of the evaluation process may involve asking industry partners for an evaluation of the TTO's services. One HEI studied by Scott *et al.* (2001), for example, hired an outside firm to measure customer satisfaction with its TTO.

The extent and operation of evaluation procedures will be contingent on each HEI and the resources on hand. Evidence suggests that even minor efforts to evaluate activity can prove beneficial – indicating to TTO staff the areas they should be emphasising, projecting an image of professionalism and transparency and, ultimately, benefiting the technology transfer process.

Lessons & Key Issues for HEIs

This chapter has been rigorous and extensive in bringing together the issues central to developing a strategic approach to technology transfer. Based on **Figure 7.1**, key issues considered included activities, structures, staffing and resources, policies and procedures, mechanisms and evaluation of technology transfer. The key factors for success and lessons that can be drawn from the discussion in each of these issues are documented in **Table 7.10**. Central to this is each HEI taking a strategic perspective – a long-term perspective that is definable, acceptable, sustainable, feasible and is continuously communicated clearly to all stakeholders.

Table 7.10: Developing a Strategic Approach to Technology Transfer: Key Issues & Facilitating Factors

Key Issue	Key Factors for Success	Facilitating Factors
1 **Strategic Management of Technology Transfer**	Development of a mission statement, strategic plan, and strategic priorities	Link objectives with the HEI's strategic plan Demonstrate how activities contribute in achieving aspects of the HEI's Mission Top leadership commitment Widespread communication of objectives Stakeholder input into objectives Acceptance and agreement of objectives by multiple stakeholders Mission statement that refers to the broad benefits of commercialisation for the public good Title of Office that reflects the chosen strategic priorities Executive Level Committees
	Focus on action, as opposed to facilitation	Assess and prioritise activities in relation to strategic objectives Build on clusters of knowledge
	Identification of technology platforms	Communicate elements of focus to business Target research funding Actively monitor research activities
2 **Technology Transfer Activities**	Active networking	Allocate time for attendance at conferences etc Databases of contacts and research Horizontal networking with other Institutions Membership of Associations/Societies
	Focus on core activities and development of core competencies	Shed non core activities Focus on getting the deal done
	Boundary-spanning	Relationship management Clear communication Database to track/manage Inventions Outreach and education

Key Issue	Key Factors for Success	Facilitating Factors
	There is no one best way to structure activities	Ensure clear lines of reporting and responsibility
		TTO must have prominent position in HEI structure and among top management
		Improve communication and teamwork among HEI personnel
		Co-ordinate the efforts of various offices to support researchers
		Close interaction between commercialisation and research activity
		Typically, TTO Directors report to Vice President/Vice Rector for Research
		Matrix structures best for co-coordinating and incentivising HEI-wide interface with business
		Although research and technology transfer are related, the two offices should be administratively independent
		Executive Committees to oversee commercialisation activities
		Co-location of offices to encourage teamwork
		Structure by specialism (activity or case manager in specific research area)
3 Structuring Technology Transfer	Shared services	Embrace innovative and creative mechanisms to overcome deficiencies
	Meso-level structures to facilitate co-operation	Complement on the ground presence with shared services to overcome resource deficiency (especially related to patenting and legislative issues)
		Shared services could include licensing, negotiation, market research for new technologies, IP management, legal expertise, spin-out creation
		HEI-based services could include raising awareness of IP, negotiating collaborative research contracts, local contacts
		Initiatives should be supported financially by government agencies. Agencies may also provide expertise to help manage and co-ordinate the structure
		Volunteer TTOs in each region should agree themselves how to set up and shape the services, and the role that each institution should play
		Strict policies regarding the conduct of member institutions particularly in terms the ownership of IP and the confidentiality of information

Key Issue	Key Factors for Success	Facilitating Factors
4 **Staffing, Skills & Resources**	Strategic staffing and continuous training and development Integrate HR strategy of TTO with strategic objectives	Need close links with researchers - facilitated by academic credentials Diverse skill base of TTO staff - understand science and willingness to put business case around it Ability to make 'cold calls' Good people management skills Leadership is crucial Accreditation and professional development standards Case mangers should be trained in a particular area of science and trained in IP management licies that nurture and reward commercialisation activities rformance appraisal of staff against objectives duction policy cement of staff/internship in other TTOs
5 **Policies & Procedures**	Requirement for user-friendly and clearly-documented IP policies	clusion of policies on video, CD-ROM or pocket-sized summaries searcher kits for each agreement visory committee made up of stakeholders to facilitate development nely responses to invention disclosures mmunicating clearly with researchers e of templates/boiler-plates or master contracts to negotiate agreements blication delays carefully monitored to preserve academic freedom

Key Issue	Key Factors for Success	Facilitating Factors
	Intellectual property rights	Clear ownership policies Policies for the distribution of royalties, preferably weighted positively towards the inventor Harmonisation in terms of national policy on IP Flexibility in application of IP rights between HEIs and business Protocol for research collaborations HEIs should not be overly-aggressive exercising their IP rights Policies for resolving conflicts of interest Keep costs in check
6 **Mechanisms for Technology Transfer**	Spin-offs and start-ups	Provide clear and transparent guidelines Pro-actively provide information, contacts and support Jointly discuss IP and decide at the outset with the academic inventors or prospective entrepreneurs whether setting up a company is the best option Prospective entrepreneurs should be allowed to remain on a part-time research or teaching position with the HEI, if they so desire
	Risk of spin-outs	Danger of too many unsustainable, low-quality spin-outs Need to assess the likelihood of success, the resource requirement and the time-line Develop expertise in this area, or else the activity becomes unsustainable and a higher risk for the HEI
	Incentive structure for spin-outs	HEI-managed incubator facilities Provision of seed funding in return for equity Focused efforts to attract management talent and catalyse company formation
	Training/seminars for prospective academic entrepreneurs	Training courses and information seminars in enterprise development for researchers

Key Issue	Key Factors for Success	Facilitating Factors
	Licensing	IP should be taken out selectively Develop capacity and resources for managing IP Proof-of-concept funding (average level of investment is up to €70,000 per invention), in conjunction with development agencies and other TTOs
	Portfolio approach: pursing many deals	TTO pursues a number of commercialisation channels to minimise the risk exposure AUTM established positive correlation between royalty income and the number of active licenses at an academic institution Need to prioritise activities (1 in 9 deals pays out royalties) More realistic in a resource-constrained environment
	Networking	To evaluate disclosures Facilitating commercialisation opportunities Formal and informal
7 **Evaluation** **Procedures**	Develop a holistic approach to evaluating technology transfer activities	Not merely performance metrics but promoted for way they facilitate/support the management of activities Focus on commercialisation outputs and TTO processes and activities Use of a variety of indicators Deal-flow is critical Link indicators to strategic priorities Top management support User involvement in development of metrics Acknowledge 'variety of excellence' Sufficient resources allocated to collection and analysing Continuous review of best practice Annual Review/Report to communicate results

8
MAKING TECHNOLOGY TRANSFER A REALITY

INTRODUCTION

In the previous chapter we documented the key issues relevant to the development of a strategic approach to technology transfer (see **Figure 7.1**). A crucial aspect of this diagram was to acknowledge that such changes cannot take place without due consideration to the motivation and awareness of researchers and a supporting ethos for commercialisation in terms of the broader context of the university. In this chapter we address the critical issues in creating a culture conducive to commercialization and the pivotal role played by researcher commitment, motivation and awareness. We conclude by relating these issues to an Irish context and highlighting the key technology transfer issues and challenges that Ireland must address if it is to succeed in making the rhetoric of becoming a knowledge based economy a reality.

CULTURE & ETHOS FOR COMMERCIALISATION

… to support technology transfer, a change in the culture and mind-set of researchers is needed.
OECD (2003: 17)

The cultural divide between researchers and business has often been acknowledged as a major impediment to knowledge creation (Leydesdorff, 2000). Technology transfer and its central objectives cannot take place without a supportive cultural context. While the impact of the complexities of commercialisation on the HEI's mission has been highlighted previously, efforts should be made to support commercialisation activities for those who wish to avail of them.

Key Success Factor 1:
A strong HEI culture of research, innovation and entrepreneurship.

In order for HEI staff and researchers to embrace fully the concept of commercialisation, there needs to be a change of mind-set. Companies and HEIs are not natural partners: their cultures and their missions are different. Academics value their freedom and independence. An over-arching aim for the TTO should be to reduce these differences by highlighting the mutual benefits to be gained from commercialisation activities and nurturing trust and common expectations between partners. As noted by Beesley (2003), 'it is apparent that a major cultural shift is necessary if collaborative networks are, first, to be established and, second, to reach their full potential'.

A cultural change will be effective, only if it is introduced at all levels of institutions. Crucially, any development of consensus and emphasis on values that respect and appreciate commercialisation will require top-down leadership from the Presidents and Deans of HEIs. Such visible commitment by top management to promote greater encouragement of collaboration between enterprise and academia is a key to success and mutual learning. The appreciation of commercial activities as being complementary to the traditional teaching and research activities of HEIs will evolve gradually overtime. As the OECD (2003: 17) noted: 'regulations are not sufficient; what is required is a gradual change of mind-set and an appreciation of the value of such activities in enhancing the role of universities in society and embedding their positive effects'.

Consensus & Mutual Understanding

Often the potential exists for commercialisation, but it is undermined by a lack of mutual understanding (Leydesdorff, 2000). Most recent research (for example, Siegel *et al.*, 2003; Friedman & Silberman, 2003; Lambert, 2003) has highlighted that there is a lack of consensus within HEIs with regard to the endorsement of commercialisation as a valuable outlet for research results. Efforts, therefore, need to be made to develop a consensus on research commercialisation. Academics and commercial entities need to understand the role of commercialisation, its relation to the HEI's mission and its benefit for the public good. As part of attempts to develop a consensus on the benefits of commercialisation, stakeholders should be consulted and informed, in order to create support and buy-in. The University of California, for example, holds retreats for academics and college administrators to deliberate and brainstorm on the topic of research commercialisation and its policies and support systems for this area of activity. Such exercises have proven to be worthwhile, as a number of amendments have been made to the University of California's policies and services as a result.

Another mechanism to promote a change in mind-set may be to locate researchers alongside technology companies. This has been found to have positive impact on research commercialisation and also sends out an implicit message of the HEI's intention. The Turku Technology Centre represents a good example of this. This mechanism of co-location creates an environment with increased opportunity for collaboration. Crucial in initiating cultural change initiatives is the dissemination of success stories and the benefits that have arisen from the process.

Role models have been instrumental in encouraging commercialisation also – both for intellectual property exploitation and spin-off enterprises. In Trinity College, Dublin, Professor Corish's collaboration with Elan in the development of the nicotine patch and Iona Technologies' spin-off example are useful role models for generating greater awareness of research commercialisation. These cases of good practice assist in generating positive images of entrepreneurship within a research environment. Success stories can be promoted through both media and workshops, with talks by speakers from campus companies and IPR 'champions' (Grace, 2003).

Formal Recognition

Another mechanism is the formal recognition of role models by the institution through awards such as President's medals or nominations for excellence in research. For example, a scheme could be established to recognise researchers who have demonstrated good practice in research commercialisation, technology transfer and contribution to industrial development and public service generally, as is the case at the University of Maryland. Such award schemes should run parallel to existing award schemes for excellence in teaching and research.

The National Development Plan notes that 'the promotion of greater collaboration within and between the RTI "supply" side (colleges, research organisations and agencies) and the "demand" side of industry has benefits for partners as well as the development of national capability in general' (NDP, 2000-06: 130). The potential researcher benefits of greater interaction with industry should be actively promoted. Collaborative ventures have been highlighted by researchers as a valuable means of identifying research ideas with commercial or industrial application. The literature highlights that the level of technology transfer has increased in environments where there is greater interaction with industry on an on-going basis. There are a number of mechanisms that can be used to increase such interaction, including the use of industry advisory groups to provide advice and feedback to individual research groups.

In a broader HEI context, the creation of an entrepreneurial climate can be linked to the curricula in place at both undergraduate and postgraduate level, which should include general entrepreneurship and innovation management training, as well as industrial application and research commercialisation. It is particularly important for science students to develop entrepreneurial skills to allow them to exploit their innovations and to develop the commercial potential of their work.

Table 8.1: The Entrepreneurship Centre at Imperial College, London

The Entrepreneurship Centre at Imperial College, London was launched in September 2000 with £2m of Science Enterprise Centre funding from the UK Government. This has been supplemented by £500,000 raised in corporate sponsorship from a range of business partners. The aim of the Entrepreneurship Centre is to embed entrepreneurship within the culture of the university and to provide faculty and students with the skills to take technical ideas to market. It provides core courses in entrepreneurship to final year students studying a wide range of subjects including medicine, engineering and science, as well as postgraduate MBA, MSc and PhD students.

A supportive cultural context is fundamental to technology transfer initiatives. Technology transfer occurs when HEI faculty and representatives from business and industry work together for mutual gain. Industry-university collaboration cannot be forced and cultural differences must be understood and attempts made at consensus and understanding (Fassin, 2000: 33). Most of the literature in this area argues that greatest challenge in terms of developing TTOs will be a change of mind-set in terms of the appropriateness and importance of commercialisation related activities in the context of university's overall mission. A summary of the factors that may facilitate in altering mind-sets and value appreciation is presented in **Table 8.2**.

By developing awareness, culture and an ethos for commercialisation, the academic researcher will be more willing to undertake research commercialisation. Key facilitating factors include the establishment of a consensus approach to research commercialisation, the provision of information, and dissemination and publicity of role models and research commercialisation success stories to both staff and students to motivate future creative innovators. Further, interrelated issues with regard to the motivation of researchers are discussed in the following section.

Table 8.2: Factors Facilitating a Change in Mind-sets

- Establish an atmosphere that promotes research among faculty.
- Leadership from the top levels.
- Encourage research directors to promote awareness of technology transfer among researchers.
- Make the TTO visible to researchers.
- Develop technology transfer policies *via* a participatory process (use faculty input to draft policies).
- Develop policies and procedures that are easy to understand.
- Interact with researchers on a one-to-one basis.
- Develop a royalty structure that allows the inventor some flexibility.
- Cultivate university/industry peer relationships with networking opportunities.
- Keep the inventor informed and in the loop during commercialization activities
- Guide the inventor: be willing to discuss whether it is worth carrying research further for commercialization purposes.
- Share successful cases with researchers.
- Clearly relate activities to the HEI mission.

RESEARCHER MOTIVATION

Key Success Factor 2:
Researcher motivation and commitment.

Personal Motivation

The most common stimulant to commercialisation of research, as identified by the majority of research studies, is the personal motivation of the individual researcher or academic in a research centre. Invention disclosures are the crucial input into the technology transfer process and cannot be simply extracted from researchers but rather must be submitted as a voluntary act. This personal motivation often stems from the background of the researcher in terms of their training and professional experience. Researchers who have previous industry experience will find it easier to collaborate and work with businesses. Furthermore, researchers who have worked in the US educational environment tend to have a greater understanding of the patenting process. In a number of cases, an important element of the personal motivation of researchers to commercialise their innovations is seeing a product derived from their innovation on sale to the public or adopted by industry.

Researcher motivation is crucial, as the number of disclosures will depend, to some extent, on the efforts of the TTO to elicit disclosures and faculty interest in technology transfer (Siegel *et al.*, 2003). Evidence shows that there is considerable heterogeneity across HEIs in their ability to elicit invention disclosures from their faculty. Interacting with the institution's TTO takes time and may not have obvious benefits for a faculty member. There is some evidence that a significant number of inventions still go unreported and unpatented (Hall, 2004). A number of activities can be seen to provide researcher motivation and facilitate in increasing the number of inventions disclosed by researchers, as discussed below.

Executive Management

The Lambert Report (2003) noted the importance of executive management within HEIs. The report referred to the importance of well-defined lines of responsibility, clearly-delegated authority and cohesive management teams of academics and administrators. This approach allows for dynamic management in an environment where decisions cannot wait for the next committee meeting. Crucially, this need not be at the expense of collegiality. Some HEIs have developed executive level committees to support the TTO, in particular, with larger, riskier commercialisation decisions. The support provided by such executive management can help promote the TTO as a professional and efficient mechanism to researchers for the exploitation of their research.

Commercialisation & Business Skills

Often, researchers are not motivated to pursue commercialisation activities simply because they are unaware of the benefits that exist or the process of commercialisation broadly. As discussed in the previous section, science undergraduate and postgraduate programmes should include business modules and courses on technology management, management principles and enterprise development to promote these issues. These courses should also address 'soft' business skills, such as oral communications, report writing and presentation skills. Science Enterprise Centres from the UK provide a good template. Their mission is to stimulate scientific entrepreneurship by teaching relevant skills to science and technology students, and helping students and staff to develop the skills required to establish and sustain start-ups based on innovative ideas.

Key Success Factor 3:
HR strategies that promote, recognise and reward
commercialisation activities.

Successful technology transfer depends, above all, on the interest and enthusiasm that faculty scientists bring to the joint research effort. Promoting such motivation generally falls under the auspices of general university policy. Thus, HEI policy can be indicative of the level of support and recognition for commercialisation activities generally (Scott *et al.*, 2001). The development and implementation of successful human resource strategies are among the most important tasks facing universities. Human resource strategies should promote and support the desired employee role behaviours among HEI staff. While human resource strategies used to support commercialisation activities cannot be introduced in all institutions immediately, they can be gradually introduced in recognition of the increasing importance of such activities. The integration of commercialisation activities into human resource strategies may become manifest in terms of allowing more leave of absence for staff, reduced workloads or simply in terms of training courses in IPR. Research by Siegel *et al.* (2003) found that HR management and other organisational practices that influence such incentives could explain some of the variation in technology transfer performance across universities. The main HR strategy levers that may provide motivation to researchers include orientation programmes, reward mechanisms and sabbaticals for researchers in industry.

Orientation Programmes

An introductory, and extremely necessary, step for successful development of technology transfer is that components of induction models should include sections on commercialisation of research, management of IP and HEI policies with regard to such issues. Indeed, in some HEIs, it is policy that a condition of taking up employment is that new staff members participate in an orientation programme, which includes basic training modules in intellectual property rights, technology transfer and the institution's support structures for academic entrepreneurship. Induction modules also provide an opportunity for communicating how such activities link with the objectives of the universities' overall strategic plan. While communication to incoming staff will generate a collective understanding, training modules could also be set up to educate existing staff on the benefits of commercialisation and university policies in place. Permanent senior researchers often undertake a more intensive programme. Both YISSUM and the University of California have excellent policies for training and educating both incoming and existing staff.

Reward Mechanisms

Generally, the current merit system in research institutions does not explicitly value commercialisation. As is evident from research in the US, rules on ownership in themselves are not sufficient; there must be incentives for institutions and for researchers to protect and exploit IP resulting from public research (OECD, 2004). In order to encourage employees and increase motivation, HEI reward systems could be modified to be consistent with technology transfer objectives. Siegel *et al.* (2003) note that 'the propensity of faculty members to disclose inventions, and thus increase the supply of technologies available for commercialisation will be related to promotion and tenure policies and the university's royalty and equity distribution formula'. Typically, researchers involved in commercialisation activities are financially rewarded through a scheme operated by the TTO – for example, a percentage of income generated from royalties and licenses. Most universities have developed a formula for this that reflects their underlying values and priorities. Many institutions allow research staff to generate considerable income for motivational purposes. In many cases, such income is also reverted to an institutional or departmental development fund, often for research purposes. This, in turn, creates strong incentive for research teams, since it produces new opportunities for further research work (Hernes & Martin, 2000). Yet, while incentives are provided through the TTO for researchers to partake in research, these incentives do often not exist at the HEI level. This presents a dilemma for researchers and can serve to de-motivate them from conducting commercialisation-related activities, as they view these as detrimental to those activities valued by universities to advance their careers (Goldfarb & Henrekson, 2002: 642).

There is widespread belief that there are insufficient rewards for faculty involvement in technology transfer (Allan, 2001). There is also hesitancy and resistance to suggested change in the human resource strategies of universities and the behaviours they nourish and support. These stem from fears that nurturing commercialisation activities may divert the direction of enquiry and draw attention away from more traditional university activities. Incentives for technology transfer, therefore, should be promoted as a complement to existing policies. HEIs must become more flexible in their reward systems for those who choose to engage in collaborative arrangements of scientific enquiry (Beesley, 2003: 1529). The strategic objectives of the HEI usually suggest support for these types of motivating mechanisms.

Researchers who engage in commercialisation activities, therefore, should be afforded the same prestige through collaborative and trans-disciplinary work as they do within their existing disciplinary structures. Current research structures have given rise to the idiom 'publish or perish' and are in direct conflict with collaborative research. Reward systems, if universities are to fully embrace of trilateral arrangements of science, need to focus on the rigor and of outcomes, rather than the field of research or the ability to publish. Often, reward structures are inconsistent with the organisational objective of increasing of the role of technology transfer as an inherent part of accomplishing the HEI mission. Well-managed staff promotion procedures that reward success in research, therefore, are an important stimulus to good and sustained performance, just as poorly-managed or arbitrary procedures that seem to prefer age and seniority, to talent and success, are important demotivators (Shattock, 2001).

Hiring, tenure, and promotion processes should give appropriate credit to university researchers who collaborate with industry. HEIs, therefore, should review and examine incentives and rewards provided to administration, faculty and staff for participating in technology transfer. Some universities have even introduced incentives specifically for university technology managers, although often HEI policy can constrain incentives in this area, unless the TTO is set up as a separate legal entity. Performance appraisal may aid in communicating the variables and objectives that technology transfer managers should be emphasising (see Simmons, 1994). Evaluation of commercialisation for a merit system should include activities such as promoting research partnerships, the application of research results and technology transfer. The University of Washington and Wayne State University are examples of universities that have recently instituted incentive compensation programmes in the TTO (Siegel *et al.*, 2003).

Further reward mechanisms operating in other HEIs include sharing of the campus portion of the license income, for education and research purposes, at a broad range of levels. Intrinsic recognition in the form of presentation of awards honouring inventors and entrepreneurs can serve as excellent motivating factors. The purpose of such human resource strategy initiatives is to institutionalise a commercial ethos. Caution must be exercised, however, that incentive systems do not encourage actors to accumulate solely quantifiable activities to the neglect of quality and value (OECD, 2003: 14).

Ultimately, research collaborations must be based on the willingness and enthusiastic participation of individual faculty members. Researchers should not be penalised for taking the commercialisation route but rather support should permeate into the human resource strategies of HEIs. The more progressive steps of redefining scholarship to include technology transfer or considering participation in commercialisation as an element in tenure or promotion decisions are evident in the stances adopted at Ohio State University and Pennsylvania State University. For most HEIs, a more realistic introductory step may be to ensure that support for commercialisation activity is clearly stated. The critical importance of this is emphasised by research by Siegel *et al.* (2003), which concluded that 'the most critical organisational factors influencing technology transfer are likely to be reward systems for faculty, TTO staffing and compensation practices, as well as actions taken by administrators to reduce informational and cultural barriers between universities and firms'.

Sabbaticals in Industry for Researchers

Research institutions should actively encourage researchers to take sabbatical leave to work in an industry for a specified time. The terms and conditions of sabbatical leave should be flexible enough to foster academic entrepreneurship. This tends to facilitate technology transfer to the company *via* the researcher's know-how. In return, the institution will enhance their linkages with the company and the researcher will develop better insights into the industry's requirements and practical aspects of achieving technology transfer. Other flexible terms of employment may also be important. MIT, for example, has clear policies allowing staff to earn funds above the normal salary through working for industry through consultancy work.

Develop Full-Time Research Career Opportunities in HEIs

Another factor that can encourage research commercialisation activities is the opportunity for permanent full-time research career paths for post-doctoral researchers who wish to stay in an academic research environment, but who do not wish to pursue a traditional academic career, which normally involves teaching. Research suggests that it is these types of academics, given adequate time following on from post-doctoral research, who are perhaps most pre-disposed to commercialisation activities and developing linkages with industry (Graff *et al.*, 2001).

Incentives & Flexibility in Employment Conditions to Stimulate Spin-off Activity

One of the barriers outlined in **Chapter 5** related the lack of time available for academics to undertake commercialisation, particularly through formation of spin-off companies. The creation of flexibility in researcher contracts is imperative. One option is a reduction of teaching hours and supervision duties, which should also include a corresponding drop in income (on a sliding scale), to afford the time to undertake commercialisation. This will provide an additional benefit to the HEI, apart from monetary reward, of new perspectives to teaching from the academic entrepreneur. It is evident from previous research and interviews that a good model for a successful academic spin-off is where the academic remains in academia and a former student works as managing director to grow the company and make it an independent business venture.

Another option is to allow academics to move from research to start spin-offs, or work with companies implementing their technologies, and then return to their research post. At some HEIs, this has been extended to include post-graduates with research ideas that have commercial potential in the form of spin-off enterprises. The Temporal Entrepreneurial Placements at the University of Twente allows post-graduates to be given positions as research assistants while they develop their business ideas and is worthy of serious consideration.

LESSONS FOR HEI LEADERS & ADMINISTRATORS

The lessons in terms of developing a culture conducive to technology transfer that motivates and provides incentives for researchers are summarised in **Table 8.3**.

MAKING TECHNOLOGY TRANSFER A REALITY IN
IRELAND

Traditionally, Ireland's export base in high-technology areas has been as a result of licensed-in technologies. Further success and growth, however, will depend on Ireland's ability to transfer the knowledge generated domestically into goods and services for world markets. Indeed, a key feature of knowledge-based economies is their ability to convert knowledge from the research base into products for economic and social benefit (Etzkowitz & Leydesdorff, 1999: 121). The National Development Plan acknowledges that 'the extent to which higher education institutions are engaged in research and development activities has a key role in determining the status and quality of these institutions and the contributions which they make to economic and social development' (National Development Plan, 2000: 129-30). Ultimately, success depends on an effective technology transfer process from the third-level sector and collaborative research between industry and academia (ESG, 2004; Etzkowitz & Leydesdorff, 1999: 121).To date, however, technology transfer mechanisms and HEI-industry collaboration in Ireland have been noted only for their *weakness*, or even *absence*. Such weaknesses have been highlighted repeatedly in recent reports – for example, Forfás (2004); Working Group Report & Recommendations (2004); Enterprise Strategy Group (2004).

Table 8.3: Creating a Culture & Ethos of Commercialisation

Key Issue	Key Factors for Success	Facilitating Factors
Culture and ethos for commercialisation	A strong university culture of research, innovation and entrepreneurship	Highlight mutual benefits to be gained from commercialisation activities Clearly relate activities to the university mission Top Level support and commitment to commercialisation Develop a consensus on research commercialisation Develop policies via a participatory process Locate researchers along side technology companies Promote and communicate success stories Formal recognition for commercialisation activities Entrepreneurship and innovation management training Undergraduate/postgraduate curricula including entrepreneurship etc Make the ILO visible to researchers
	Motivated and entrepreneurial researchers	Keep the inventor informed and in the loop during commercialisation activities User friendly policies Relationship management Clear communication Database to track/manage Inventions Outreach and education
Researcher motivation	HR strategies that recognise and reward commercialisation	Orientation programmes with IP modules Reward Mechanisms that recognize commercialisation activities Inclusion of commercialisation activities in tenure and promotion decisions Sabbaticals for researchers in industry Full-time research opportunities Permanent/Contract ratio alignment Incentives and flexibility in employment conditions to stimulate spin-off activity

Developing Technology Transfer in Ireland

While the importance of a strategic approach to technology transfer has been highlighted throughout this book, there are a number of points with particular relevance to the Irish context that are worth examining further.

Communication & Awareness

In picking up the key theme of this chapter, successful technology transfer can only take place in the context of an environment that is conducive to commercialisation activities. Ireland's deficiencies in this respect are well-documented (see Forfás, 2004; 2004a). The ultimate challenge for all Irish HEIs in developing third-stream activities is to make academic entrepreneurship a reality at an institutional level. A critical impasse in terms of developing technology transfer stems from the lack of awareness or commercialisation experience among academics (Forfás, 2004a). King's (2005) study[30] of 1,000 research academics, for example, demonstrates that Irish HEIs have a lot of ground to make up. Half of the respondent researchers in his sample have never been involved in collaborative research with industry, while 70% of the respondents had never been involved in any commercialisation activity resulting from academic research. The validity and role of commercialisation activities in terms of delivering research findings for the public good should be communicated and highlighted among Irish HEIs as an inherent part of the HEI's mission (Den Hertog et al., 2003). It should be stressed that commercialisation activities *complement*, not *cannibalise*, existing HEI activities. Indeed, the reality is that all HEI activities are not mutually exclusive. Practically, the promotion of technology transfer as a valuable outlet for research results is probably best achieved by putting in place support and incentive systems, while balancing these activities with prudent governance structures.

Procedures & Policies

Even where a consensus might exist as to the value of technology transfer, efforts are often impeded by poor information and confusion concerning the arrangements for developing and exploiting IP. In the US, clarity was provided through the introduction of the Bayh-Dole Act 1980. In an Irish context, the implementation of ICSTI's *National Code of Practice* (April 2004) will go some way in addressing this issue, although it is crucial for Irish HEIs to develop clear and well-documented policies and practices that can be communicated to researchers in a user-friendly manner.

---

[30]   As reported in Downes, J. (2005).

Technology Platforms

Given Ireland's relatively weak position in terms of resources available to compete on a global basis, it is critical that focused effort is directed in developing core competencies in specific technologies where we have advantages and expertise. Research that draws on special capabilities in Ireland's academic and industrial system (for example, agri-food, cell cycle control, enabling technologies, medical biotechnology, biopharmaceuticals) should be promoted actively. Together, HEIs and the Irish government need to target research investments in areas with the most likely scientific and economic impact, and where the country already has concentrated skills and industrial interests, such as computers, electronics, pharmaceuticals and medical equipment (Harris, 2005: 25). Typically, the development of such 'technology platforms' will be interdisciplinary in nature, built on knowledge from a number of specific domains. Ireland's ability to create this coherent focus, while at the same time leveraging diverse ranges of knowledge, will be critical if it is to retain its economic momentum.

Shared Services Facility

The reality facing all Irish HEIs is they will not be in a position to provide all the services required for HEI-industry collaboration. HEIs, therefore, will have to examine the best means of co-operation with other HEIs, regionally and nationally, in order to achieve their long-term objectives. Shared service facilities or meso-level structures may facilitate co-operation among TTOs, thereby reducing resource and expert deficiencies and leveraging economies of scale (see **Chapter 7**). The OECD (2002) has identified that several countries that are already experimenting with regional or sector-based offices – for example, Denmark, UK, and Germany. Evidently, smaller HEIs, such as those in Ireland, can benefit more either by aligning themselves with the TTOs of larger universities or by forming collective TTOs among themselves in order to capture economies of scale. In Ireland, several inter-institutional alliances are already in place, such as the Atlantic University Alliance, the Technology Transfer Initiative and TecNet. These alliances have evolved around collaboration in research and educational programmes (Forfás, 2004a: 35).

The idea of some level of shared costs and regional alliances in the Ireland, therefore, seems to be sensible. A national, publicly-funded *shared services facility* would allow HEIs to achieve globally-competitive economies of scale in the management of intellectual property. There are strong arguments for providing Irish HEIs with access to pooled expertise in a number of resource-intensive areas, such as patent applications, licensing, market research, venture capital, commercial partnerships, asset management, staff training and international promotion. Examples of successful university commercialisation networks include the multi-campus University of California and the multi-university Connect Midlands, led by Warwick University (see **Table 8.4**).

Irish HEIs should position themselves at the forefront of these path-breaking developments in Ireland's national innovation system and cluster strategy. Further, the EU is placing increased emphasis on collaborative mechanisms, although pan-European initiatives to date have been limited (an example is the European Science Foundation's EUROCORES programme, a research scheme combining national and European financial resources to support European scientists in addressing major research challenges).

Recently, there have been proposals to develop a Connect Europe, to serve as a transnational network for technology transfer and commercial development. The network would bring together emerging technology businesses with the resources they need to succeed, with a particular focus on improving access to finance. Currently, there are already a number of European Connect organisations already in existence, such as Connect Denmark, Connect Estonia, and Connect Sweden. In the USA, the Association of University Technology Managers also serves as a useful template for the development of a national or European association with similar objectives.

Portfolio Approach to Technology Transfer

Patents or other forms of IP should be taken out selectively, as not all results need to be appropriated or should remain in the public domain. Ultimately, IP policy should be about balancing support for commercialisation of publicly-funded research and entrepreneurship at HEIs, on one hand, and the protection of public access to IP generated with public research funds on the other (OECD, 2003: 104). The Lambert Report (2003) exercised caution in over-emphasising the virtues of company formation, noting that high spin-out rates may come at a cost to licensing. Irish HEIs are probably best advised to take a portfolio approach to technology transfer, in order to spread the risk of commercialisation activities.

Table 8.4: Overview of Connect Midlands

Connect Midlands, an organisation established by the University of Warwick in the British Midlands in 2001, is the premier network of choice in the Midlands region for technology companies seeking access to finance. From a standing start, over the last 3.5 years, Connect have achieved an impressive track record:

♦ 450 active and networking members (Entrepreneurs, Researchers, Investors, Mentors)

♦ 185 entrepreneur presentations so far to investors from the UK and international investors

♦ € 44m investment raised by 64 companies

♦ Additional € 5.8m of offers

♦ 168 jobs created/safeguarded

The prime purpose of Connect is to act as a facilitator and networking vehicle, bringing together emerging technology businesses with individuals and organisations from the capital, technology, corporate, research and business support communities. Connect Midlands has a portfolio of development programmes to support technology entrepreneurs including the Connect InvoRed programme. Connect InvoRed is € 4.8m investment-readiness programme, helping over 350 companies get ready to raise investment. Connect also deliver:

♦ Series of funding platforms where companies make their pitch to the investment community. Over the last 3 years, Connect has established a reputable track record raising over 52 million euros with a further 8 million euros in the pipeline.

♦ Events such as 'Meet the Entrepreneur' and 'Enterprise Workshops' to help entrepreneurs develop skills and knowledge to create and manage high-growth companies and share best practice. Connect support the technology-transfer activities of over 10 of its regional university members.

♦ Workshops such as 'Technology Briefings' and 'Bootcamps', amongst others, to promote the latest R&D expertise originating from research and nurturing collaborative opportunities with the private sector in commercialization and technology-transfer.

Source: www.connectmidlands.org.

TOWARDS A STRATEGIC APPROACH TO
TECHNOLOGY TRANSFER IN IRELAND

*The production of primary research information is not the end but the
beginning of a process that continues until the usefulness of that
information is realised. The commercialisation of research and
knowledge for Ireland's economic benefits through effective intellectual
property management and technology transfer needs to be a priority for
all higher education and public research institutes and it is essential that
institutes establish strong capabilities in this regard.*
Forfás (2004: 29).

The propensity to commercialise, as well the development of
knowledge, and realistic expectations as to the nature of the
technology transfer process, are best developed by Technology
Transfer Officers working on the ground and developing long-term
relationships with academics. Such efforts necessarily should take
place in the context of HEIs developing a clearer focus to their mission
and objectives, as well as the policies to implement these. The Lambert
Review (2003) in the UK highlighted the need for effective leadership
and the encouragement of skills among HEI leaders and
administrators. In the light of these challenges, Irish HEI leaders and
administrators have to think, act and manage strategically. Short term
initiatives or maintaining the *status quo* are no longer an option if HEIs
are to play their part in the knowledge-based economy. In developing
a strategic perspective, HEIs will need to consider which third-stream
activities their institutions will pursue, given their current areas of
international expertise and knowledge competencies.

Strategic management of technology transfer involves setting a
long-term direction, having assessed the internal and external
environment of the HEI, and understood multiple stakeholder
perspectives. Objectives, policies and incentives, in turn, should flow
from strategic priorities. In assessing the merits of various strategic
priorities for the HEI, it is useful to consider:

♦ **Suitability:** Does the strategy address the key issues identified in the
   strategic analysis? Useful for *screening* choices (often qualitative)

♦ **Feasibility:** Can the strategy be made to work in practice? Focus on
   practicalities of resourcing

♦ **Acceptability:** Are the expected outcomes *acceptable*? (To *whom*?)

HEIs individually, and Ireland as a nation, needs to develop a coherent
technology transfer strategy at both a national and regional level. Such
efforts are critical as 'national innovation systems cannot be expected

to thrive where the culture of institutional and organizational fragmentation prevails' (Malairaja & Zawdie, 2004).

Government & Macro-level Support

Development of a more strategic approach to technology transfer is necessarily contingent on government support and resources. It is only in the context of macro-investment and government support that Ireland can rectify its deficiencies in key innovation indicators (for example, patents, researchers per thousand labour force). Further, given that they provide the raw material for the commercialisation process, effort needs to be made to increase dramatically the number of active researchers. Recent media accounts of the low intake into science and engineering degrees, despite government efforts in this area, are not encouraging. Fortunately, these challenges and the potential of technology transfer and commercialisation activities have not gone unnoticed. Harris (2005: 24-25) notes 'in fact, across all government departments, there is an impressive commitment to policies, programmes, and investment designed to make Ireland an enduring knowledge society'. This commitment is probably best evidenced by introduction of a National Science Advisor, who will report to a Cabinet-level committee dedicated to science. Such commitment has to be sustained if Ireland is to prosper, both economically and socially, from technology transfer initiatives.

Risks, Rewards & Balance

It is important to acknowledge that the commercialisation process involves both risks and rewards, and that only a small proportion of HEIs internationally have succeeded in earning a significant percentage of their total research expenditure in licensing revenues or returns on equity. On the other hand, the resources required for investment in the process, at both national and institutional levels, are substantial. Expectations, therefore, need to be managed carefully. While revenue generation must be one element of the overall rationale for commercialisation – to be undertaken as efficiently and effectively as possible – others include long-term national and regional economic development, the creation of high skill graduate employment, the growth of Irish research capacity as part of the national innovation system and, in the context of the Lisbon agenda, the development of Europe's knowledge-based economy and society. These elements are themselves a strong justification for increased public funding, not only of the university-based research process as such, but also the structures, personnel and physical space needed for the successful implementation of technology transfer and commercialisation.

Further, it is important to note that there are multiple channels and methods, direct and indirect, formal and informal, for technology transfer. Thus, while the commercialisation of property owned by universities is an important component of third-stream activities, it is only one among the many other functions that link HEIs with society (Molas-Gollart *et al.*, 2002). There are certain firms, especially within the pharmaceutical and software industry, that have a considerable interest in co-operating continuously with researchers at the university but, for most firms, the most important link has to do with the recruitment of well-educated graduates. This implies, of course, that it would be dubious to completely overhaul the organisational design and regulatory frameworks of HEIs with reference to the new, quite exceptional, tendencies in the research fields of biotechnology and the life sciences. Again, the focus should be on complementing rather than cannibalising existing activities. In acknowledgement of this, while supportive of the triple helix concept, we favour an interpretation of the triple helix as one that emphasises interaction, rather than intersection, or interdependence, between HEIs and business. For Irish HEIs, the key challenge is one of balancing the tension between maintaining high teaching standards with developing sustainable collaborative efforts, while remaining true to their scholastic traditions.

In pursuing an approach that recognises the 'varieties of excellence' at best practice institutions, we appreciate that there is no 'one best way' to structure and pursue technology transfer initiatives. Each HEI is a product of a distinct process of social, economic and intellectual development, and finds its own balance between teaching, research and a wide set of third-stream activities (Molas-Gollart *et al.*, 2002: 9). Attempts that seek simply to replicate or generalise from country-specific cases therefore are futile. In practice, technology transfer will be conditioned by a number of factors, including, for example, the institutional setting and history of the HEI, the commitment of the HEI's leadership and previous experience in technology transfer (Sheehan & Wyckoff, 2003: 37). Our contention is that, underpinning most best practice cases, however, is a strategic approach to technology transfer that recognises the value of clear policies and procedures, researcher commitment and awareness and appreciates the interdependence between structures, activities, expertise and the technology transfer mechanisms pursued. This, we believe, is the key lesson for Irish HEIs as they pursue their agenda of turning the rhetoric of becoming a knowledge-based economy into a reality.

Appendix 1
ACRONYMS & ABBREVIATIONS

AUTM	Association of University Technology Managers
BERD	Business Expenditure on Research & Development
EI	Enterprise Ireland
EPO	European Patent Office
ESG	Enterprise Strategy Group
HEA	Higher Education Authority
HEI	Higher Education Institution
HERD	Research & Development in the Higher Education Sector
HLEG	High-level Expert Group
ICSTI	Irish Council for Science, Technology & Innovation
ICT	Information and communication technologies
ILO	Industrial Liaison Office
IP	Intellectual Property
IPO	Irish Patents Office
IPR	Intellectual Property Rights
NDP	National Development Plan
NSF US	National Science Foundation
OECD	Organisation for Economic Co-operation & Development
PRTLI	Programme for Research in Third Level Institutions
R&D	Research and Development
SFI	Science Foundation Ireland
TTO	Technology Transfer Office
USPTO	United States Patent & Trademark Office

Appendix 2
USEFUL SOURCES OF INFORMATION

Journals/ Working Papers
American Economic Review
DRUID Working Paper Series
Journal of the Association of University Technology Managers
Journal of Research Administration
Journal of Technology Transfer
MIT Sloan Management Review
NBER Working Paper Series
Public Administration Review
R&D Management
Research Policy
Science & Public Policy
Technovation

Key Policy Reports

Enterprise Strategy Group (2004). *Ahead of the Curve: Ireland's Place in the Global Economy: Report of the Enterprise Strategy Group*, Dublin: Enterprise Strategy Group.

Forfás (2004). *Building Ireland's Knowledge Economy- The Irish Action Plan for promoting in R&D to 2010*, Dublin: Forfás.

Forfás (2005). *Science Foundation Ireland – the First Five Years 2001-2005: Report of an International Evaluation Panel*, Dublin: Forfás.

ICSTI (2004). *National Code of Practice for Managing Intellectual Property from Publicly-funded Research*, Dublin: ICSTI.

Lambert, R. (2003). *Lambert Review of Business-University Collaboration: Final Report*, London: HM Treasury.

OECD (2003). *Turning Business into Science: Patenting & Licensing at Public Research Organisations*, Paris: OECD.

Web Resources – Agencies/Organisations

http: //www.autm.net
http: //www.connectmidlands.org
http: //www.enterprise-ireland.com
http: //www.forfas.ie
http: //www.forfas.ie/icsti
http: //www.hea.ie
http: //www.ictireland.ie
http: //www.intertradeireland.com/
http: //www.expertiseireland.com
http: //www.irchss.ie/http: //www.ircset.ie
http: //www.lesi.org
http: //www.sfi.ie
http: //www.sussex.ac.uk/spru
http: //www.tecnet.ie
http: //www.unesco.org
http: //www.cambridge-mit.org
http: //www.cisc.ie

Web Resources – Useful Information

http: //www.cordis.lu/praxis
http: //www.business.auc.dk/druid/wp/wp.html
http: //les.man.ac.uk/cric/papers.htm
http: //www.rand.org/
http: //www.isi.fhg.de/homeisi.htm
http: //www.step.no/
http: //www.sussex.ac.uk/spru
http: //meritbbs.rulimburg.nl/rmpdf/rmlist.html

Appendix 3
BIBLIOGRAPHY

Allan, M. (2001). 'A review of best practices in university technology licensing offices', *The Journal of the Association of University Technology Managers*, Volume XIII.

Arnold, E., Bussillet, S., Swoden, P., Stroyan, P., Whitehouse, S. & Zaman, R. (2004). *Technopolis: An Evaluation for the RTDI for Collaboration Programme: Main Report*, Dublin: Forfás.

Association of University Technology Managers (2002). *Licensing Survey FY 2002 – Survey Summary*, Northbrook, IL: Association of University Technology Managers (http: //wwwautm.net).

Association of University Technology Managers (2003). *Licensing Survey FY 2003 – Survey Summary*, Northbrook, IL: Association of University Technology Managers (http: //wwwautm.net).

AUTM – *see* Association of University Technology Managers.

Balthasar, A., Battig, C., Thierstein, A. & Wilhelm, B. (2000). 'Developers: key actors of the innovation process. Types of developers and their contacts to institutions involved in research and development, continuing education and training, and the transfer of technology', *Technovation*, Vol.20, pp.523-38.

Beesley, L. (2003). 'Science policy in changing times: Are governments posed to take full advantage of an institution in transition?', *Research Policy*, Vol.32. pp.1519-31.

Bercovitz, J., Feldmean, M., Feller, I. & Burton, R. (2001). 'Organisational structure as determinant of academic patent and licensing behaviour: An exploratory study of Duke, John Hopkins, and Pennsylvania State Universities', *Journal of Technology Transfer*, Vol.26(1-2), pp.21-35.

BHEF – *see* Business Higher Education Forum.

Blumenstyk, G. (1995). 'Turning off spin-offs: Bucking a trend, University of Arizona ends direct commercialising of faculty research', *Chronicle of Higher Education*, 21 July, p.A33.

Bok, D. (2003). *Universities in the Marketplace: The Commercialisation of Higher Education*, Princeton, NJ: Princeton University Press.

Bozeman, B. & Coker, K. (1992). 'Assessing the effectiveness of technology transfer from US government R&D laboratories: The impact of market orientation', *Technovation*, Vol.12(4), pp.239-55.

Bozeman, B. (2000). 'Technology transfer and public policy: A review of research and theory, *Research Policy*, Vol.29, pp.627-55.

Brennan, J. & Shah, T. (2003). *Managing Quality in Higher Education: An International Perspective on Institutional Assessment & Change,* Buckingham: The Society for Research into Higher Education.

Brint, S. (2005). 'Creating the future: "New Directions" in American research universities', *Minerva,* Vol.43, pp.23-50.

Burns, T. & Stalker, G.M. (1961). *The Management of Innovation,* 2nd edition, London: Tavistock Publications.

Business Higher Education Forum (2003). *Working Together Creating Knowledge: The University-Industry Collaboration Initiative,* Washington, DC: Business Higher Education Forum (http: //www.bhef.com/includes/pdf/working-together.pdf).

Carayannis, E. & Alexander, J. (1999). 'Winning by co-opeting in strategic government-university-industry R&D partnerships: The power of complex, dynamic knowledge networks', *Journal of Technology Transfer,* Vol.24(2-3), pp.197-210.

Carlsson, B. & Fridh, A.C. (2002). 'Technology transfer in the United States' universities', *Journal of Evolutionary Economics,* Vol.12, pp.199-232.

Castillo, F., Parker, D., & Zilberman, D. (2000). 'Offices of technology transfer and privatisation of university discourse', *Working Paper,* University of California, Berkeley: Department of Agricultural & Resource Economics.

Chen, S. (2005). 'Extending internationalisation theory: A new perspective on international technology transfer', *Journal of International Business Studies,* Vol.36, pp.231-45.

Chesborough, H. (2003). *Open Innovation,* Boston, MA: Harvard Business School Press.

Chrisman, J., Hynes, T. & Fraser, S. (1995). 'Faculty entrepreneurship and economic development: The case of the University of Calgary', *Journal of Business Venturing,* Vol.10 (4), pp.267-81.

Clarke, B. (1998). *Creating Entrepreneurial Universities: Organisation Pathways of Transformation,* Oxford: International Association of Universities & Elsevier Science.

Connelan, L. (2004). 'Ireland needs to double number of researchers', *Sunday Business Post,* 23 May.

Cope, G. (2004). 'Developing a knowledge-based economy', *Initiative,* December/January, pp.12-3.

Coupe, T. (2003). 'Science is golden: Academic R&D and university patents', *Journal of Technology Transfer,* Vol.28(1), pp.31-45.

Cozzens, S., Healey, P., *et al.* (1990). *The Research System in Transition,* Boston, MA: Kluwer Academic Publishers.

Cronin, M. (2004). 'Chief Executive's Report', in *Forfás Annual Report 2004,* Dublin: Forfás.

Cunningham, J. & O'Gorman, C. (2001). *Enterprise in Action: An Introduction to Entrepreneurship in an Irish Context,* 2nd edition, Cork: Oak Tree Press.

Dean, C. & LeMaster, J. (1995). 'Present barriers to technology transfer: US to Eastern Europe *versus* US to Mexico', *The International Executive*, Vol.33, pp.35-42.

Den Hertog, P. & Roelandt, T. (1999). 'Cluster analysis & policy-making: The state of the art' in Roelandt, T. & Den Hertog, P. (ed.) (1999), *Boosting Innovation: The Cluster Approach*, OECD Proceedings, Paris: OECD, pp.413-27.

Den Hertog, P., Gering, T. & Cervantes, M. (2003). 'Introduction and Overview', in OECD (2003). *Turning Business into Science: Patenting & Licensing at Public Research Organisations'*, Paris: OECD.

Department of Trade & Industry (2004). *White Paper on Science & Technology*, London: HMSO.

Di Gregorio, D. & Shane, S. (2003). 'Why do some universities generate more start-ups than others?', *Research Policy*, Vol.32, pp.209-27.

Downes, J. (2005). Academics 'fail to commercialise research'. *The Irish Times*, 10 August.

DTI – *see* Department of Trade & Industry.

EIMS – *see* European Innovation Monitoring System.

Enterprise Strategy Group (2004). *Ahead of the Curve: Ireland's Place in the Global Economy: Report of the Enterprise Strategy Group'*, Dublin: Enterprise Strategy Group.

ESG – *see* Enterprise Strategy Group.

Etzkowitz, H. & Leydesdorff, L. (1997). *Universities & the Global Knowledge Economy: A Triple Helix of University-Industry-Government Relations*, London: Cassell Academic.

Etzkowitz, H. & Leydesdorff, L. (1999). 'The future location of research and technology transfer', *Journal of Technology Transfer*, Vol.24(2-3), pp.111-23.

Etzkowitz, H. & Leydesdorff, L. (2000). 'The dynamics of innovation: From national systems and Mode 2 to a triple helix of university-industry-government relations', *Research Policy*, Vol.19, pp.109-23.

Etzkowitz, H. (2002). *MIT & the Rise of Entrepreneurial Science*, London/New York: Routledge.

Etzkowitz, H. (2003) 'Research groups as 'quasi-firms': the invention of the entrepreneurial university', *Research Policy*, Vol.32(1), pp.109-21.

European Commission (2003). *European Trend Chart on Innovation: Country report, Finland* (http: //trendchart.cordis.lu).

European Innovation Monitoring System (1995). *Good Practice in the Transfer of University Technology to Industry: Case Studies by a Consortium Led by Innov GmbH*, EIMS Publication, No.26 (http: //www.cordis.lu/eims/src/eims-r26.htm).

Fassin (2001) – referenced on page 43, 123, 183

Fassin, Y. (1991). 'Academic ethos *versus* business ethics', *International Journal of Technology Management*, Vol.6(5/6), pp.533-46.

Fassin, Y. (2000). 'The strategic role of university Industry Liaison Offices', *Journal of Research Administration*, Vol.1(2), pp.31-41.

Feldman, M., Feller, I., Bercovitz, J. & Burton, R. (2002). 'Equity and the technology transfer strategies of American research universities', *Management Science*, Vol.48(1), pp.99, 105-22.

Forfás (2003). *Research & Development in Ireland 2001- At a Glance*, Dublin: Forfás.

Forfás (2004). *Building Ireland's Knowledge Economy: The Irish Action Plan for Promoting Investment in R&D to 2010'*, Report to the Inter-Departmental Committee on Science & Technology Innovation, Dublin: Forfás.

Forfás (2004a). *From Research to the Marketplace: Patent Registration & Technology Transfer in Ireland*, Dublin: Forfás.

Forfás (2004b). *Annual Report*, Dublin: Forfás.

Forfás (2005). *Business Expenditure on R&D*, Dublin: Forfás Science & Technology Indicators Unit.

Fraiman, N. (2002). 'Building relationships between university and business: The case of Columbia Business School', *Interfaces*, Vol.32(2), March/April.

Friedman, J. & Silberman, S. (2003). 'University technology transfer: Do incentives, management and location matter?', *Journal of Technology Transfer*, Vol.28(1), pp.17-30.

Friedman, T.L. (2005). 'Follow the Leapin' Leprechaun', *The New York Times*, 1 July.

Georghiou, L. (2004). 'Evaluation of behavioural additionality: A concept paper', paper presented at the European Conference on Good Practice in Research Evaluation & Indicators, *'Research and the Knowledge Based Society: Measuring the link'*, 24 May, NUI, Galway.

Gering, T. & Schmoch, U. (2003). 'The management of intellectual assets by German public research organisations', Chapter 9 in OECD (2003), *Turning Business into Science: Patenting & Licensing at Public Research Organisations*, Paris: OECD, pp.169-88.

Gibbons, M., Limoges, C., Nowotny, H., Schwartzman, S., Scott, P. & Trow, M. (1994). *The New Production of Knowledge: The Dynamics of Science & Research in Contemporary Societies*, London: Sage Publications.

Godin, B. & Gingras, Y. (2000). 'The place of universities in the system of knowledge production', *Research Policy*, Vol.29(2), pp.273-78.

Godin, B. (2005). *The Linear Model of Innovation: The Historical Construction of an Analytical Framework*, Turin: Triple Helix 5.

Goldfarb, B. & Henrekson, M. (2002). 'Bottom-up *versus* top-down policies towards the commercialisation of university intellectual property', *Research Policy*, Vol.32, pp.639-58.

Goldhor, R.S. & Lund, R. (1983). 'University to industry advanced technology transfer: A case study', *Research Policy*, Vol.12(3), pp.121-52.

Grady, R. & Pratt, J. (2000). 'UK technology transfer system calls for stronger links between higher education and industry', *Journal of Technology Transfer*, Vol.25(2), pp.205-11.

Graff, G., Heiman, A., Zilberman, D., Castillo, F. & Parker, D. (2001). 'Universities, technology transfer and industrial R&D' (http://are.berkely.edu/~ggraff/Graff-et-al-University-TT.pdf).

Graff, G., Heimen, A. & Zilberman, D. (2002). 'University research and Offices of Technology Transfer', *Californian Management Review*, Vol.45(1), pp.88-115.

Gulbrandsen, M. & Etzkowitz, H. (1999). 'Convergence between Europe and America: The transition from industrial to innovation policy', *Journal of Technology Transfer*, Vol.23(2-3), pp.223-33.

Hall, B. (2004). 'University-industry research partnerships in the US', paper presented at Kansai Conference, February 2004 (available online at http://emlab.berkeley.edu/pub/users/bhhall/papers/BHH%20IP-Univ-Ind.pdf).

Harmon, B., Alexander, A. & Cardozo, R. (1997). 'Mapping the university technology transfer process', *Journal of Business Venturing*, Vol.12, pp.423-34.

Harris, W. (2005). 'Secrets of the Celtic Tiger: Act Two', *Issues in Science & Technology*, Summer, pp.23-27.

Healy, C. & Buckley, P. (2005). 'Ireland: A global growth leader', *Fortune*, 1 August, Special Advertising Section, S1-S10.

Henderson, R., Jaffe, A. & Trajtenberg, M. (1998). 'Universities as a source of commercial technology: A detailed analysis of university patenting, 1965-88', *Review of Economics & Statistics*, Vol.80(1), pp.119-27.

Hernes, G. & Martin, M. (2000). *Management of University-Industry Linkages*, Policy Forum No.11, Paris: UNESCO/International Institute for Educational Planning (http: //unesdoc.unesco.org/images/0012/001235/123538e.pdf)

Higher Level Expert Group (2005). *Frontier Research: The European Challenge*, Luxembourg: Office for Official Publications of the European Communities, European Commission.

HLEG – *see* Higher Level Expert Group.

Howells, J. & MacKinlay, C. (1999). *Commercialisation of University research in Europe: Report to the Expert Panel on the Commercialisation of University Research for Advisory Council on Science & Technology*, Ontario: Advisory Council on Science & Technology.

ICSTI (2004). *National Code of Practice for Managing Intellectual Property from Publicly-funded Research*, Dublin: ICSTI.

ICT Ireland (2004). '*Commercialisation of R&D in the ICT Sector: Working Group Report & Recommendations*, Dublin: ICT Ireland.

ICT Ireland (2004a). *Creating Europe's Most Attractive Environment for Intellectual Property'*, Submission by the Irish Academy of Engineering, the Institution of Engineers of Ireland and ICT Ireland to Science Foundation Ireland, Dublin: ICT Ireland.

IMD – *see* International Institute for Management Development.

International Institute for Management Development (2000). *The World Competitiveness Yearbook 2000*, Lausanne: IMD International.

Jain, R. (2004). 'Editorial', Special Edition on Economic & Environmental Issues, *International Journal of Technology Transfer & Commercialisation*, Vol.4(3), pp.253-55.

Jankowski, J. (1999). 'Trends in academic research spending, alliances and commercialisation', *Journal of Technology Transfer*, Vol.24(1), pp.55-68.

Jensen, R. & Thursby, M. (2001). 'Proofs and prototypes for sale: The licensing of university inventions', *American Economic Review*, Vol.91(1), pp.240-59.

Jones-Evans, D. (1999) Universities, technology transfer &spin-off companies: Academic entrepreneurship in different European regions', *Targeted Socio-Economic Research Project No.1042*, Pontypridd: University of Glamorgan Business School.

Jones-Evans, D., Klofsten, M., Andersson, E. & Pandya, D. (1999). 'Creating a bridge between university and industry in small European countries: The role of the Industrial Liaison Office', *R&D Management*, Vol.29(1), pp.47-56.

Karlsson, M. (2004). *Research Commercialisation in the US*, City?: Swedish Institute for Growth Policy Studies.

Kaukonen, E. & Nieminen, M. (1999). 'Modelling the triple helix from a small country perspective: The case of Finland', *Journal of Technology Transfer*, Vol.24(2-3), pp.173-83.

Kline, S. & Rosenberg, N. (1986). 'An overview of innovation', in Landau, R. & Rosenberg, N. (Eds.). *The Positive Sum Game*, Washington DC, National Academy Press.

Klofsten, M. & Jones-Evans, D. (2000). 'Comparing academic entrepreneurship in Europe – The case of Sweden & Ireland, Small Business Economics, Vol.14, pp.299-309.

Lach, S. & Shankerman, M. (2003). *Incentives in Universities*, CEPR Discussion Paper 3916 (http://www.cepr.org/pubs/dps/DP3916.asp).

Lambert, R. (2003). *Lambert Review of Business-University Collaboration: Final Report*, London: HM Treasury.

Large, D., Belinko, K. & Kalligatsi, K. (2000). 'Building successful technology commercialisation teams: Pilot empirical support for the theory of cascading commitment', *Journal of Technology Transfer*, Vol.25(2), pp.169-80.

Larson, K. (2000). 'In response to Moshe Vigdor's paper', in Hernes, G. & Martin, M. (2000). *Management of University-Industry Linkages*, Policy Forum No.11, Paris: UNESCO/International Institute for Educational Planning.

Lee, Y. (2000). 'The sustainability of university-industry research collaboration: An empirical assessment', *Journal of Technology Transfer*, Vol.25(2), pp.111.

Leydesdorff, L. & Etzkowitz, H. (2001). 'The transformation of university-industry-government relations', *Electronic Journal of Sociology*, Vol.5(4).

Leydesdorff, L. (2000). 'The triple helix: An evolutionary model of innovations', *Research Policy*, Vol.29, pp.243-55.

Leydesdorff, L. A. & Etzkowitz, H. (1996). 'The triple helix as a model for innovation studies', *Science & Public Policy*, Vol.25(3), pp.195-203.

Lindberg, C. (2000). 'In response to Guilherme Ary Plonski', in Hernes, G. & Martin, M. (2000). *Management of University-Industry Linkages*, Policy

Forum No.11, Paris: UNESCO/International Institute for Educational Planning.

Lundvall, B.-A. (1998). 'Why study national systems and national styles of innovation?', *Technology Analysis & Strategic Management*, Vol.10(4), pp.407-21.

Lundvall, B.-A. (2002). 'The university in the learning economy', *DRUID Working Paper No.02:06*, Copenhagen: Copenhagen Business School.

Malairaja, C. & Zawdie, G. (2004). 'The "black box" syndrome in technology transfer and the challenge of innovation in developing countries: The case of international joint ventures in Malaysia', *International Journal of Technology Management & Sustainable Development*, Vol.3(3), p.233-51.

Malerba, F. (2004). 'Sectoral systems: How and why innovation differs across sectors' in Faberberg, J., Mowery, D. & Nelson, R.R. *The Oxford Handbook of Innovation*, Oxford: Oxford University Press.

Mansfield, E. (1991). 'Academic research and industrial innovation', *Research Policy*, Vol.20, pp.1-12.

Mason, R. (2003). 'The university: Current challenges and opportunities', in D'Antoni, S. (Ed.). *The Virtual University: Models & Messages, Lessons from Case Studies*, Paris: UNESCO/International Institute for Educational Planning.

Melara, C. & Arcelus, M. (2005). 'The role of industry for promoting research-based spin-outs from university and other research institutions', Paper AD167 (http://www.triplehelix5.com/pdf/A167_THC5.pdf).

Meseri, O. & Maital, S. (2001). 'A survey analysis of university technology transfer in Israel: An evaluation of projects and determinants of success', *Journal of Technology Transfer*, Vol.26(1), pp.115-27.

Meyer, J.F. (2003). *Strengthening the Technology Capabilities of Louisiana Universities*, Report for the Louisiana Department of Economic Development, Louisiana: Pappas & Associates.

Mitra, J. & Formica, P. (Eds.) (1997). *Innovation & Economic Development, University-Enterprise Partnership in Action*, Dublin: Oak Tree Press.

Molas-Gollart, J., Salter, A., Patel, P., Scott, A. & Duran, X. (2002). 'Measuring third-stream activities: Final report to the Russell Group of Universities', Brighton: SPRU Science & Technology Policy Research.

Mowery, D. & Arvids, A.Z. (2002). 'Academic patent quality and quantity before and after the Bayh-Dole Act in the United States', *Research Policy*, Vol.31(3), pp.399-418.

Mowery, D., Nelson, C., Sampsat, B. & Ziedonis, A. (2001). 'The growth of patenting and licensing by US universities: An assessment of the effects of the Bayh-Dole Act of 1980', *Research Policy*, Vol.30, pp.99-119.

Mowery, D., Nelson, R., Sampat, B. & Arvids, A.Z. (2004). *Ivory Tower & Industrial Innovation: US University-Industry Technology Transfer before and after the Bayh-Dole Act*, Stanford, CA: Stanford University Press.

Mowery, D., Rosenberg, N. & Landau, R. (1992). *Technology & the Wealth of Nations*, Stanford, CA: Stanford University Press.

National Development Plan (2000). *Ireland: National Development Plan 2000-2006*, Dublin: Government Publications.

National Health Service (1998). *The Management of Intellectual Property & Related Matters: An Introductory Handbook for R&D Managers & Advisors in NHS Trusts & Independent Providers of NHS Services*, London: National Health Service.

National Institute of Standards & Technology (2002). *A Toolkit for Evaluating Public R&D Investment Models & Methods* (http://www.atp.nist.gov).

National Science Foundation (2003). *Science & Engineering Indicators 2001*, Arlington, VA: National Science Foundation.

National University of Ireland, Galway (2002). Strategic Plan 2003-2008, Galway, National University of Ireland, Galway.

NDP – *see* National Development Plan.

Nedeva, M., Georghiou, L. & Halfpenny, P. (1999), 'Benefactor or beneficiary: The role of industry in the support of university research equipment, *Journal of Technology Transfer*, Vol.24(2/3), pp.111-23.

Nelson, L. (1998). 'The rise of intellectual property protection in the American university', *Science*, Vol.270(5356), pp.1460-61.

Nelson, L. (2001). 'University technology transfer practices: Reconciling the academic and commercial interests in data access and use' (http: //www.nap.edu/html/codata_2nd/ch13.html).

Nelson, R. (1994). 'Science-technology-economy interactions', in Granstrand, O. (ed.), *Economics of Technology*, Amsterdam: Elsevier Science, pp.323-38.

Nelson, R. R. (1993). *National Systems of Innovation: A Comparative Study*, Oxford: Oxford University Press.

Newton, D. & Henricks, M. (2003). 'Can entrepreneurship be taught?', *Entrepreneur*, April.

NHS – see National Health Service.

NIST – see National Institute of Standards & Technology.

NSF – see National Science Foundation.

OECD (1994). *Assessing & Expanding the Science & Technology Base*, Paris: OECD.

OECD (2001). *Innovation & the Strategic Use of IPRs*, paper DSTI/STP/TIP (2001) 4, Paris: OECD.

OECD (2002). *Interim Results of the TIP Project on the Strategic Use of IPRs at PROs*, Working Document, Paris: OECD.

OECD (2003). *Turning Business into Science: Patenting & Licensing at Public Research Organisations*, Paris: OECD.

OECD (2004). *Report on OECD-China Events on Intellectual Property Rights held in Beijing, China* (http://www.oecd.org/dataoecd/46/10/32267101.pdf).

PACEC – *see* Public & Corporate Economic Consultants.

Padmanabhan, V. & Souder, W.E. (1994). 'A Brownian motion model for technology transfer: Application to a machine maintenance expert system', *Journal of Product Innovation Management*, Vol.11(2), pp.119-33.

Pandya, D. & Cunningham, J. (2000). *A Review with respect to the Commercialisation of Non-Commissioned Research in Ireland*, report to Commercialisation of Research Taskforce, ICSTI, Dublin: Irish Council for Science Technology & Innovation.

Parker, D. & Zilberman, D. (1993). 'University technology transfers: Impacts on local & US economies', *Contemporary Policy Issues*, Vol.11(2), pp.87-99.

Piekarski, A.E.T. & Torkomian, A.L.V. (2005). *How R&D Public Financing Incites the Academy-Industry Cooperation: An Assessment of the Effects of Public Policy in Brazil*, Turin: Triple Helix 5.

Plonski, G. (2000). 'In what way do changing university-industry relations affect academic activities within higher education institutions', in Hernes, G. & Martin, M. (2000). *Management of University-Industry Linkages*, Policy Forum No.11, Paris: UNESCO/International Institute for Educational Planning.

Porter, M.E. & Stern, S. (2001). 'Innovation: Location matters', *MIT Sloan Management Review*, Vol.42(4), pp.28-36.

Porter, M.E. (1990). *The Competitive Advantage of Nations*, New York: Free Press.

Postlewait, A., Douglas, P. & Zilberman, D. (1993). 'The advent of biotechnology and technology transfer in agriculture', *Technological Forecasting & Social Change*, Vol.43, pp.271-87.

Powell, W.W., Koput, K.W. & Smith-Doerr, L. (1996). 'Inter-organisational collaboration & the locus of innovation: Networks of learning in bio-technology', *Administrative Science Quarterly*, Vol.41, pp.116-45.

Public & Corporate Economic Consultants (2003). *The Cambridge Phenomenon: Fulfilling the Potential: Main Report*, Cambridge: Public & Corporate Economic Consultants (http://gcp.uk.net/SITE/UPLOAD/DOCUMENT/GCP_Main_Report.pdf).

Rahm, D. (1994). 'Academic perceptions of university-firm technology transfer', *Policy Studies Journal*, Vol.22(2), pp.267-78.

Rahm, D., Bozeman, B. & Crow, M. (1998). 'Domestic technology transfer & competitiveness: An empirical assessment of roles of university and governmental R&D laboratories', *Public Administration Review*, Vol.48, pp.969-78.

Rip, A. & VanderMeulen, B. (1996). 'The post-modern research system', *Science & Public Policy*, Vol.23(6), pp.343-52.

Roessner, J.D. (1996). 'University-Industry Collaborations: Choose the Right Metric'. *Science's Next Wave*, June.

Rogers, E.M., Yin, J. & Hoffman, J. (2000). 'Assessing the effectiveness of technology transfer offices at US research universities', *Journal of the Association of University Technology Managers*, Vol.12, pp.47-80.

Rosenberg, N. & Nelson, R. (1994). 'American university and technical advance in industry', *Research Policy*, Vol.23, pp.323-48.

Sanchez, A. & Tejedor, A. (1995). 'University-industry relationships in peripheral regions: The case of Aragon in Spain', *Technovation*, Vol.15(10), pp.61.

Schindel, D.E. (2002). *United States Survey Response,* OECD/DTSI Steering & Funding Public Research Organisations project, Paris: OECD.

Scott, A., Steyn, G., Geuna, A., Brusoni, S. & Steinmueller, E. (2001). *The Economic Returns to Basic Research & the Benefits of University-Industry Relationships: A Literature Review and Update of Findings,* Report for the Office of Science & Technology, Brighton: SPRU - Science & Technology Policy Research (http://www.sussex.ac.uk/spru/documents/review_for_ost_final.pdf).

Shane, S. (2003). 'Executive forum: University technology transfer to entrepreneurial companies', *Journal of Business Venturing,* Vol.17(6), pp.537-52.

Shattock, M.L. (2001). 'In what way do changing university-industry relations affect the management of higher education institutions?' in Hernes, G. & Martin, M. (Eds.) (2001). *Management of University-Industry Linkages',* Policy Forum No.11, Paris: UNESCO/International Institute for Educational Planning.

Sheehan, J. & Wyckoff, A. (2003). *Targeting R&D: Economic & Policy Implications of Increasing R&D Spending,* STI Working Paper 2003/8, Paris: Science & Innovation Unit, OECD.

Siegel, D., Waldman, D., & Link, A. (2003). 'Assessing the impact of organisational practices on the productivity of university technology transfer offices: An exploratory study', *Research Policy,* Vol.32, pp.27-48.

Simons, R. (1994). 'How new top managers use control systems as levers of strategic renewal', *Strategic Management Journal,* Vol.15, pp.169-89.

Smith, H. (1991). 'Industry-academic links: The case of Oxford University', *Environment & Planning,* Vol.9(5), pp.403-16.

Stankiewicz, R. (1986). *Academics & Entrepreneurs: Developing University-Industry Relations,* London: Frances Pinter Publications.

Stevenson, A. (2003). 'Twenty years of academic licensing: Royalty income and economic impact', *Licensing Executive Society Journal,* September, pp.133-40.

Stokes, D. E. (1997). *Pasteur's Quadrant: Basic Science & Technological Innovation,* Washington DC: Brookings Press.

Thursby, J. & Kemp, S. (1999). 'Growth & productive efficiency of university intellectual property licensing', *Research Policy,* Vol.28(1), pp.109-24.

Thursby, J. & Kemp, S. (2001). 'Who is selling the ivory tower? Sources of growth in university licensing', *Management Science,* Vol.48, pp.90-104.

Thursby, J., Jensen, R. & Thursby, M. (2001). 'Objectives, characteristics & outcomes of university licensing: A survey of major US universities', *Journal of Technology Transfer,* Vol.26, No.1(2), pp.59-72.

Tijssen, R. (2001). Global and domestic utilization of industrial relevant science: patent citation analysis of science-technology interactions and knowledge flows', *Research Policy,* Vol.30, pp.35-54.

UK University Commercialisation Survey (2003). *UK University Commercialisation Survey FY 2002,* Nottingham: Auril / Nottingham Business School / UNICO.

United States Patent & Trademark Office (2003). *Performance &*
   Accountability Indicators Report FY 2003, Alexandria, VA: United States
   Patent & Trademark Office (http://uspto.gov/web/offices/com/
   annual/index.html)
University of Arizona Office of Economic Development (2004). *Technology*
   Transfer at the University of Arizona: A Comparative Analysis &
   Benchmarking Study, Tucson, AZ: University of Arizona.
US Congress (1985). *Hearings of the Committee on Science & Technology, US*
   House of Representatives, 98th Congress, Second Session, 21 March 1984,
   Washington, DC: United States Government Printing Office.
USPTO – see United States Patent & Trademark Office.
Van Dierdonck, R. & Debackere, K. (1988). 'Academic entrepreneurship at
   Belgian universities', *R&D Management*, Vol.18(4), pp.341-53.
World Economic Forum (2001). *Global Competitiveness Report 2001-2002*,
   Oxford: Oxford University Press.
Ziman, J. (1994). *Prometheus Bound: Science in a Dynamic Steady State*,
   Cambridge: Cambridge University Press.

APPENDIX 4
FURTHER READING

Argyres, N. & Liebeskind, J. (1988). 'Privatising the intellectual commons: Universities and the commercialisation of biotechnology', *Journal of Economic Behaviour & Organisation*, Vol.35, pp.427-54.

Bush, V. (1945). *Science: The Endless Frontier*, Washington: United States Government Printing Office.

Cohen, W., Florida, R., & Goe, W. R. (1994). *University Industry Research Centers in the United States*, Pittsburgh, PA: Carnegie Mellon University.

Connolly, B. (2000). 'Research is the root of the IT tree of knowledge', *Irish Independent*, 30 March.

Feldman, M. & Desrochers, P. (2001). 'Truth for its own sake: Academic culture and technology transfer at John Hopkins University', September (http://www.cs.jhu.edu/~mfeldman/Minerva102.pdf).

Hammel, G. & Sampler, J. (1998). 'The E-Corporation', *Fortune*, 7 December, pp.52-63.

Harper, J. & Rainer, K. (2000). 'Analysis and classification of problem statements in technology transfer', *Journal of Technology Transfer*, Vol.25(2), pp.135-57.

Hicks, D., Breitzman, A., Hamilton, K. & Narin, F. (2000). 'Research excellence and patented innovation', *Science & Public Policy*, Vol.27(5), pp.310-20.

Hull, C. (1990). *Technology Transfer between Higher Education & Industry in Europe*, Luxembourg: TII.

O'Higgins, E. (2002). 'Government and the creation of the Celtic Tiger: Can management maintain the momentum?', *Academy of Management Executive*, Vol.16(3), pp.104-20.

Okubo, Y. (1997). *Bibliometric Indicators and Analysis of Research Systems: Methods and Examples*, STI Working Paper 1997/1, Paris: OECD.

Ormerod, R. (1996). Combining management consultancy and research', *International Journal of Management Science*, Vol.24(1), pp.1-12.

Peters, M. (2004). 'Universities into business: How the campuses compare', Special Report on Spin-outs, *The Observer*, 4 April.

Phillimore, J. (1999). 'Beyond the linear view of innovation in science park evaluation: An analysis of Western Australian technology parks', *Technovation*, Vol.19(11), pp.673-80.

Rahm, D., Bozeman, B. & Crow, M. (1988). 'Domestic technology transfer and competitiveness: An empirical assessment of roles of university and governmental R&D laboratories', *Public Administration Review*, Vol.48, pp.969-78.

Scherer, F.M. & Herhoff, D. (2000). 'Technology policy for a world of skewed distributed outcomes', *Research Policy*, Vol.29, pp.559-66.

Scott, S. (2004). *Academic Entrepreneurship: University Spin-offs & Wealth Creation*, Northampton: Edward Elgar.

Shane, S. (1999). 'Technology regimes and new firm formation', Working paper, Smith School of Business, University of Maryland.

Smircich, L. (1983). 'Concepts of culture and organizational analysis', *Administrative Science Quarterly*, Vol.28, pp.339-58.

Snow, C.P. (1959). *The Two Cultures*, Cambridge, MA: Cambridge University Press.

Stefanovich, G.B. (2003). *What do universities expect from the National Association of Technology Transfer Managers?*, Presentation to OECD/Ministry of Education of the Russian Federation, CDRF Workshop: Commercialising Intellectual Property, Yekaterinburg, 9 December.

Thursby, J. & Thursby, M. (2000). *Who is Selling the Ivory Tower?: Sources of Growth in University Licensing*, NBER Working Paper #7718.

Vigdor, M. (2000). 'Organisational responses: The management of interfaces. Experiences from the Hebrew University Jerusalem, Israel', in Hernes, G. & Martin, M. (Eds.) (2000). *Management of University-Industry Linkages*, Policy Forum No.11, Paris: UNESCO/International Institute for Educational Planning.

INDEX

Abo Akademi University
  Foundation (Finland)   167, 169
Aboatech Ltd (Finland)   167
Academy of Finland   164, 165-66
Akamai   146
American Cyanamid   162
AMT Ireland   70
Association of University
  Technology Managers   116, 131,
              170, 183, 199, 217, 243
Association of University-related
  Research Parks   214
Aston University (UK)   106
Atlantic University Alliance 29, 70,
                                  242
Australia
  licensing   86
  start-ups   94

B.G. Negev (Israel)   157
Bar Ilan Research & Development
  Company (Israel)   157
Bar-Ilan University (Israel)   157
Bayh-Dole Act 1980 (USA)   8, 15,
              43, 51, 112, 144, 171, 241
Belgium
  Interuniversity Institute for
  biotechnology   111, 195
  start-ups   94
  University of Louvain La Neuve
                            94-95, 216
Ben-Gurion University of Negev
  (Israel)   157
Bioresearch Ireland   70
BITS   103-104
BMBF (Germany)   111

Brazil
  University of Sau Paulo   26, 190
Brigham Young University (USA)
                                   92
Bristol University (UK)   106
business expenditure on R&D   10
  by sector   11
business incubators
  US universities   147-148

California Institute for Quantitative
  Biomedical Research   154, 171
California Institute of Technology
  (CalTech) (USA)   106
Cambridge University (UK)   71-72
Cambridge-MIT Institute
  (UK/USA)   113
Canada
  Canadian-Israel Industrial
  Research & Development
  Foundation   156
  University of British Columbia
                                  203
Canadian-Israel Industrial
  Research & Development
  Foundation   156
case management approach   27,
                              190-91
Chalmers University, Gothenburg
  (Denmark)   68, 201
Chiron   153
Ciba-Geigy   162
Colminatum Oy (Finland)   167
Columbia University (USA) 52, 82,
      90, 92, 95, 100, 134, 146, 147, 199

commercialisation, barriers to
   cultural   123-126
   empirical evidence   126-127
   institutional   118-121
   operational   121-123
commercialisation, stimulants to
   macro   110-112
   micro   112-118
company formation
   factors influencing in universities
                       96-97
   start-ups and spin-offs   92-101
   US universities   147-48
conflict of interest
   OECD definition   204
Connect Denmark   243
Connect Estonia   243
Connect Europe   243
Connect Midlands (UK)   104, 243
Connect Sweden   243
CONNECT   154-155, 171
CORDIS   185
Corish, Professor   231

Dartmouth University (USA)   92
"death of distance"   7
Denmark
  Chalmers University,
  Gothenburg   68, 201
  Connect Denmark   243
  measures of performance of
  science base   17
Dimotech (Israel)   157, 158, 160
Dublin Business Innovation Centre
                     216
Duke University (USA)   23, 32, 33,
        37-38, 42, 95, 185, 190
Durham University (UK)   72

Edinburgh College of Art
  (Scotland)   103
Edinburgh Crystal (Scotland)   103
Edinburgh University (Scotland) 72
Elan   231

Enterprise Ireland   70, 199
  Commercialisation Fund   84
  Patent Fund   80
Enterprise Strategy Group   2
entrepreneurial university   6, 15
Estonia
  Connect Estonia   243
EU   111
  CORDIS   185
  Fifth Framework programme   92
  Lisbon Agenda   121
EU-15
  measures of performance of
  science base   17
Europe
  Connect Europe   243
  gross expenditure on R&D   10
European Science Foundation   243
EUROCORES   243
EUROSTAT   136

Federal Reserve (USA)   144
Fifth Framework programme (EU)
                     92
FileMaker Pro   185, 210
Finland   164-169
  Abo Akademi University
  Foundation
                 167, 169
  Aboatech Ltd   167
  Academy of   164, 165-66
  Colminatum Oy   167
  Finnish innovation system 165-66
  Finn-Medi   168
  Foundation for Finnish
  Inventions   165, 166, 167
  Hermia   179
  INNOTULI   66
  measures of performance of
  science base   17
  Ministry of Education   165
  Ministry of Trade & Industry 165
  Nokia   169
  Oteniemi Science Park   167
  patent applications   79

SITRA – National Fund for
Research & Development   66,
                        165, 166, 167
Spinno Business Development
Centre   167
Tampere University Hospital 168
Tampere University of
Technology   168, 169
Technical University of Helsinki
                             167, 216
TEKES – National Technology
Agency   164, 165, 166, 167
Turku School of Economics &
Business Administration   169
Turku Technology Centre169, 231
University of Tampere   168
University of Turku Foundation
                                  167
University of Turku   167, 169
Finn-Medi (Finland)   168
Florida State University 52, 92, 147
Foundation for Finnish Inventions
                        165, 166, 167
"frontier research"   181

Genentech   153
Georgetown University (USA)   59,
                        178, 215-216
  strategic technology transfer 178
Germany
  BMBF   111
  Ministry of Research &
  Technology   29
  start-ups   94
Glasgow University (Scotland)   72
GlaxoSmithKline   72
"Golden Circle", The (UK)   71
Greenspan, Alan   144
gross expenditure on R&D   10

Harney, Mary – An Tánaiste   1, 78
Harvard University (USA)   52,
                             178, 200

Hebrew University of Jerusalem
  (Israel)   22, 26, 30, 143,
                156-157, 161-64, 177, 190
Hermia (Finland)   168
Higher Education Authority   179
  Programme for Research in Third
  Level Institutions   179

IBM   7, 146, 169
Imperial College London (UK)
                        71-72, 106, 199
  Entrepreneurship Centre   232
INNOTULI (Finland)   66
innovation
  and multiple actors   8
  Finnish system of   165-66
institutional equity holdings 97-98
intellectual property   77-78
intellectual property rights 77-78
Interuniversity Institute for
  Biotechnology (Belgium) 111, 195
invention disclosure   81
  criteria used to assess, in Israel157
  University of California   150
Iona Technologies   231
Ireland
  developing technology transfer in
                                 241-245
  developing strategic approach to
  technology transfer in   245-247
  measures of performance of
  science base   17
  patent applications   79
Irish Business & Employers
  Confederation   78
Irish Council for Science,
  Technology & Innovation   207
ISIS Oxford (UK)   87
Israel   155-164
  B.G. Negev   157
  Bar Ilan Research & Development
  Company   157
  Ben-Gurion University of Negev
                                  157

Canadian-Israel Industrial
Research & Development
Foundation   156
Dimotech   157, 158, 160
Hebrew University of Jerusalem
                                22, 26, 30, 143,
                156-157, 161-64, 177, 190
Ministry of Industry & Trade,
Chief Scientist's Office   159
RAD-RAMOT   161
RAMOT   157, 160-161
Technicon-Israel Institute of
Technology   157
Technion Entrepreneurial
Incubator Company 158-159, 160
Technion Research &
Development Foundation Ltd
                                        158-159
Tel Aviv University   157, 160
USA-Israel Science Foundation
                                             156
Weizman Institute   30, 157
Yeda   30, 157
YISSUM   26, 30, 157, 161-164,
                     170, 171, 190, 235
Italy
  licensing   86
  start-ups   94

Japan
  gross expenditure on R&D   10
  start-ups   94
John Hopkins University 23, 32, 34,
        36-37, 41, 100, 130, 185, 210
Johnson & Johnson   162

Kings College London (UK)   106
Knowledge Transfer Partnerships
  (UK)   105
knowledge-based economy
  HEIs in   2-3
Korea
  licensing   86
  start-ups   94

Leeds University (UK)   106
Licensing Executives Society (USA)
                                             183
licensing   84-92, 214-215
  cash and equity   88-89
  exclusivity   89
  income   90-92
  MIT   87-88
  negotiating   89-90
  OECD   86, 87
  Oxford   87-88
  revenues, US universities   147
  UK   86
Linkoping University (Sweden) 182
Lisbon Agenda   121

McKinnell, Hank   117
measures of performance of science
  base   17
Merck   162
Micro-Electronics Innovation &
  Computer Research
  Opportunities programme
  (California, USA)   211
Ministry of Education (Finland) 165
Ministry of Industry & Trade, Chief
  Scientist's Office (Israel)   159
Ministry of Research & Technology
  (Germany)   29
Ministry of Trade & Industry
  (Finland)   165
mission statement(s)   176, 178,
MIT (USA)   43, 45, 57, 66, 72, 77,
            87, 92, 97, 104, 107, 134,
            144, 145, 146, 147, 161,
            171, 192, 200, 213, 238
  licensing   87-88
  Magnet programme   161, 171

National Business Incubator
  Association (USA)   214
National Development Plan
  (Ireland)   1, 231, 239
National Science Advisor   246

Netherlands
  licensing 86
  start-ups 94
  University of Twente 99, 239
Newcastle University (UK) 72
Nokia (Finland) 169
Non-patent IP actions 101
North Carolina State University
  (USA) 57, 134, 189
Northwestern University (USA)
  52, 148, 208
Northwestern University
Evanstown Research Park (USA)
  147
Norway
  start-ups 94
NovaUCD 95
NUI Galway 8, 70, 180

OECD 136
  conflict of interest, definition 204
  technology transfer studies
  132-133
Ohio State University (USA)
  134, 237
organisational structures
  characteristsics 40-42
  co-ordination 40-42
  co-ordination capacity 41
  ILO performance &
  organisational structures 41
  incentive alignment 41
  information processing 40-42
  information-processing capacity
  41
  locus of decisions 40-42
Oteniemi Science Park (Finland)167
Oxford Brookes University 106
Oxford University (UK) 71-72,
  87-88, 100
  licensing 87-88

Pasteur's Quadrant 181
patent(s) 78-79
  applications 79

assigned 80
US universities 80, 147
Pennsylvania State University 23,
  32, 33, 38-39, 41-42, 189, 237
Pfizer 117, 162
PharmaSTART 154, 171
Pracecis Pharmaceuticals 146
Programmes in Advanced
  Technologies 70
Proof of Concept Fund (Scotland)
  83

RAD-RAMOT (Israel) 161
RAMOT (Israel) 157, 160-161
research commercialisation &
  technology transfer
  characteristics of the transfer
  agent 71
  mechanisms 68
  rational decision-making
  perspective 64
  relationship perspective 64-65
  strategies 69-70
  success factors for 69
research commercialisation
  culture 229-233, 240
  ethos 229-233, 240
  success factors 230, 233, 235
research
  business demand for excellence
  9-10
  environment in Ireland 16
  European and Irish experience
  10-11
  impetus for increased emphasis
  on commercialisation 7-9
  sponsored by industry,ownership
  of 102
Research Triangle Park (USA) 214
researcher motivation 233-239, 240
  commercialisation & business
  skills 234-235
  executive management 234
  incentives to stimulate spin-offs
  238-239
  orientation programmes 235

personal motivation 233-234
reward mechanisms 236-238
sabbaticals 238
Russia
start-ups 94

Sandoz 162
Schering 162
science enterprise centres (UK) 234
Science Foundation Ireland 55, 179
Scotland
Edinburgh College of Art 103
Edinburgh Crystal 103
Edinburgh University 72
Glasgow University 72
Proof of Concept Fund 83
Singapore
Technopreneurship 21 199
SITRA – National Fund for
Research & Development
(Finland) 66, 165, 166, 167
soft methods of technology transfer
103-107
consultancy 106-107
internships 103-104
Knowledge Transfer Partnerships
(UK) 105
networks & partnerships 104-106
Spain
start-ups 94
Spinno Business Development
Centre (Finland) 167
spin-offs
best practice for universities in
managing 98-99
Twente University 99
University of Louvain La Neuve
94-95
SRI International 154
Stanford University (USA) 15, 43,
66, 90, 92, 95, 118, 134,
144, 145, 147, 168, 171, 200
strategic technology transfer
173-228
activities 179-187
boundary-spanning 184-187

case management approach 191
co-location of offices 189
conflicts of Interest 204-206
co-operative regional/national
structure, guidelines for 196
core activities/competences 184
cost control 210-212
culture 174
degree of autonomy 189-190
developing in Ireland 245-247
disclosure process 202-203
ethos 174
evaluation 216-222
evaluation, annual report/review
222
evaluation, potential indicators
219-221
Executive Committees 188-189
Georgetown University 178
industry visits 187
integration of strategy with HR
strategy 200
intellectual property 206-208
intellectual property protocol 207
intellectual property, flexibility in
policy for 208
intellectual property, guidelines
210
leadership 197-198
licensing 214-215
mechanisms 212-216
mission statement(s) 178
negotiating agreements 203-204
networking 182-184, 215-216
organisational structure 187-196
organisational structure,
guidelines for developing 189
outreach & education 186-187
policies & procedures 201-212
portfolio approach 215-216
publishing constraints, perceived
206
regional co-operation 192-194
resource constraints, structural
options to overcome 192
resources 201-212
shared services 194-196

skilled & flexible staff   200
spin-offs   212-214
sponsored research   206
staff & resources   196-201
staff recruitment, guidelines for
                                      197
start-ups   212-216
strategic plans   176-178
structural changes required   191
structure by specialism   190-196
success factors   176, 179,
                                188, 192, 212
technology platforms   179-182
training & professional
development   199-200
training for prospective academic
entrepreneurs   214
Swansea University (UK)   106
Sweden
  Connect Sweden   243
  Linkoping University   182
  measures of performance of
  science base   17
Switzerland
  licensing   86
  start-ups   94
Sykes, Sir Richard   72

Tampere University Hospital
(Finland)   168
Tampere University of Technology
  (Finland)   168, 169
Technical University of Helsinki
  (Finland)   167, 216
Technicon-Israel Institute of
  Technology (Israel)   157
Technion Entrepreneurial
  Incubator Company (Israel)
                                158-159, 160
Technion Research & Development
  Foundation Ltd (Israel)   158-159
technological balance of payments
                                      136
Technology Transfer Initiative
                                  29, 242

Technology Transfer Office(s)
  activities, see TTO activities
  difficulty in measuring activities
                                128-130
  external support for   29
  funding   28
  historical context   43-44
  metrics   129-130
  mission statement(s)   52
  models   25-26
  optimal institutional
  arrangements for   30-31
  organisational arrangements   26
  organisational structure as
  determinant of academic patent
  & licensing behaviour   32-33
  organisational forms, see TTO
  organisational forms
  organisational structure,
  guidelines for developing   31
  other names for   23-24
  responsibilities   51-55
  responsibilities to multiple
  stakeholders   51
  role(s)   44-50
  staffing, see TTO staffing
  stakeholder characteristic(s)   54
  structures   27
  UK   24
  varieties of organisational
  structures   32-42
technology transfer   72-84
  activity indicators   136-140
  benchmarking   133
  company model   143, 167
  contingency effectiveness model
                                      140
  developing in Ireland   241-245
  indicators   134
  measuring   138-139
  mechanisms   77-84
  OECD Studies   132-133
  output & impact indicators
                                131-136
  process   72-73
  quantitative measures   132, 135

soft methods of, *see* soft methods of technology transfer
strategic, *see* strategic technology transfer
start dates of US university programmes 115
regional co-operation model 143, 167
technology
strategic approach in Ireland 245-247
Technopreneurship 21 (Singapore) 199
TecNet 29, 242
TEKES – National Technology Agency (Finland) 164, 165, 166, 167
Tel Aviv University (Israel)157, 160
third-stream activities 13-15, 175
balance 175
benefits 13-14
concerns 14-15
strategically managing 15-17
Transfert 94
TRDF
commercialization process 159
Triangle Universities Licensing Consortium (Louisiana, USA) 194
Trinity College, Dublin 216, 231
triple helix 2, 3-7, 15, 105
role of academia 6-7
role of government 6
role of industry 5
TTO activities
difficulty in measuring 128-130
management of intellectual property 44, 47
management of technology transfer 44, 45-46
network development 44, 45, 48-49
switchboard services 44-45
TTO organisational forms
H-form 23, 36-37, 41
matrix 23, 39-40, 42, 191
M-form 23, 38-39, 41-42

TTO staffing 55-60
essential characteristics of 59
functions 59-60
levels 56-57
management approach 28
personnel profile 58-59
priorities of 53
technology transfer skills 57-59
Turku School of Economics & Business Administration 169
Turku Technology Centre (Finland) 169
Turku University Hospital 167

UK
Aston University 106
Bristol University 106
Cambridge University 71-72
Cambridge-MIT Institute 113
Connect Midlands (UK) 104, 243
Durham University 72
"Golden Circle" 71
Imperial College London 71-72, 106, 199
ISIS Oxford 87
Kings College London 106
Knowledge Transfer Partnerships 105
Leeds University 106
licensing 86
Newcastle University 72
Oxford Brookes University 106
Oxford University 71-72, 87-88, 100
science enterprise centres 234
Swansea University 106
University of Manchester Institute of Science & Technology (UMIST) 72
Warwick University 28, 52, 104
Wolverhampton University 103
United States Patent & Trademark Office 112
universities
multiple roles 2
University College Cork 70

University of Arizona   214
University of British Columbia
    (Canada)   203
University of California – Berkeley
                                43
University of California   15, 22, 43,
        45, 90, 143, 144, 147, 171,
        177, 186, 193, 197, 200,
        211, 230-31, 235, 243
    invention disclosure   150
University of Georgia (USA)   26
University of Limerick   70
University of Louvain La Neuve
    (Belgium)   94-95, 216
University of Manchester Institute
    of Science & technology (UMIST)
    (UK)   72
University of Maryland (USA)   231
University of Oregon (USA)
                            147, 201
University of Sau Paulo (Brazil)
                             26, 190
University of Tampere (Finland)
                                168
University of Texas   147
University of Turku Foundation167
University of Turku   167, 169
University of Twente
    (Netherlands)   99, 239
University of Washington (USA)
                                237
University of Wisconsin (USA)   43
university-industry collaboration
                              65-72
    company formation   68
    matching services   66
    TTOs and   66-68
university-industry cultural
    differences   124
UPM Kymmene   169
USA
    Bayh-Dole Act 1980   8, 15, 43, 51,
            112, 144, 171, 241-42
    Brigham Young University   92

California Institute for
Quantitative Biomedical Research
                            154, 171
California Institute of Technology
    (CalTech)   106
Cambridge-MIT Institute   113
Columbia University52, 82, 90, 92,
        95, 100, 134, 146, 147, 199
CONNECT   154-155, 171
Dartmouth University   92
Duke University 23, 32, 33, 37-38,
            42, 95, 185, 190
Federal Reserve   144
Florida State University   52, 92,
                                147
Georgetown University   59, 178,
                            215-216
gross expenditure on R&D   10
Harvard University   52, 178, 200
indicators   134
invention disclosure   81
John Hopkins University   23, 32,
    33,   36-37, 41, 100, 130, 185, 210
Licensing Executives Society   183
licensing   86
measures of performance of
    science base   17
Micro-Electronics Innovation &
Computer Research
Opportunities programme
    (California)   211
MIT   43, 45, 57, 66, 72, 77, 87, 92,
        97, 104, 107, 134, 144, 145, 146,
        147, 161, 171, 192, 200, 213, 238
National Business Incubator
Association   214
North Carolina State University
                            57, 134, 189
Northwestern University
Evanstown Research Park   147
Northwestern University
                            52, 148, 208
Ohio State University   134, 237
patents assigned   80
Pennsylvania State University
                        23, 32, 33, 38-39,
                        41-42, 189, 237

Research Triangle Park   214
Stanford University   134, 147
start-ups   93, 94
Triangle Universities Licensing
Consortium (Louisiana)   194
universities and business
incubators   147-148
universities and company
formation   147-48
University of Arizona   214
University of California   15, 22,
            43, 45, 90, 143, 144, 147,
               171, 177, 186, 193, 197,
            200, 211, 230-31, 235, 243
University of Georgia   26
University of Maryland   231
University of Oregon   147, 201
University of Texas   147
University of Washington   237
University of Wisconsin   43
USA-Israel Science Foundation
                               156
Wayne State University   237
Yale University   66, 82, 203
USA-Israel Science Foundation 156

Warwick University UK)28, 52, 104
Wayne State University (USA)   237
Weizmann Institute (Israel) 30, 157
Wolverhampton University   103

Yale University   66, 82, 203
Yeda (Israel)   30, 157
YISSUM (Israel)   26, 30, 157,
            161-164, 170, 171, 190, 235

OAK TREE PRESS

is Ireland's leading business book publisher.

It develops and delivers
information, advice and resources
to entrepreneurs and managers –
and those who educate and support them.

Its print, software and web materials
are in use in Ireland, the UK, Finland,
Greece, Norway and Slovenia.

OAK TREE PRESS

19 Rutland Street
Cork, Ireland
T: + 353 21 4313855
F: + 353 21 4313496
E: info@oaktreepress.com
W: www.oaktreepress.com